Copyright © 2003 by Ruşen MEYLANİ.
All Rights Reserved.

All graphics and text based on TI – 83 Plus have been used with the permission granted by Texas Instruments.

This book is protected by United States and International Copyright Laws. No part of this book may be reproduced in any form, by photostat, microfilm, xerography, or any other means, or incorporated into any information retrieval system, electronic or mechanical, without the written permission of the copyright owner.

Worldwide Distributor:
Ruşen MEYLANİ and Mustafa Atakan ARIBURNU
RUSH Publications and Educational Consultancy, LLC
220 East 31st Street, Apt# 1D, Baltimore, MD
21218, USA

All inquiries, suggestions and comments should be addressed to:

Ruşen MEYLANİ
RUSH Publications and Educational Consultancy
Akatlar Zeytinoglu C, Banka Evleri, A2 Blok, Kat: 3, Daire: 7
Etiler, İstanbul
80630, TURKEY

url	:	http://www.rushsociety.com
e-mail	:	rushco@superonline.com
phone	:	+90 212 351 8924
fax	:	+90 212 351 8925

Cover design by: Pınar ERKORKMAZ; **e-mail:** pinare1@e-kolay.net
Printed in İstanbul, Turkey; by Seçil Ofset; **phone:** +90 212 629 0615

ISBN: 0 – 9748868 – 0 – 7

Edited by: Ahmet ARDUÇ

Advanced Calculation and Graphing Techniques with the TI – 83 Plus Graphing Calculator

Unauthorized copying or reuse of any part of this page is illegal.

Copyright © 2003 by Ruşen MEYLANİ. All Rights Reserved.

This book is protected by United States and International Copyright Laws. Therefore unauthorized copying or reuse of any part of this book is illegal.

All graphics and text based on TI – 83 Plus have been used with the permission granted by Texas Instruments.

From: Bassuk, Larry <l-bassuk@ti.com>
Sent: Thursday, February 27, 2002 16:23 PM
To: Meylani, Rusen <meylani@superonline.com>; Foster, Herbert <h-foster@ti.com>; Vidori, Erdel <e-vidori@ti.com>
RE: USAGE OF TI 83+ FACILITIES IN MY SAT II MATH BOOKS
Rusen Meylani,
Again we thank you for your interest in the calculators made by Texas Instruments.
Texas Instruments is pleased to grant you permission to copy graphical representations of our calculators and to copy graphics and text that describes the use of our calculators for use in the two books you mention in your e-mail below.
We ask that you provide the following credit for each representation of our calculators and the same credit, in a way that does not interrupt the flow of the book, for the copied graphics and text:
Courtesy Texas Instruments
Regards,
Larry Bassuk
Copyright Counsel
972-917-5458

-----Original Message-----
From: Bassuk, Larry
Sent: Thursday, February 21, 2002 9:14 AM
To: 'Rusen Meylani'; Foster, Herbert
Subject: RE: USAGE OF TI 83+ FACILITIES IN MY SAT II MATH BOOKS
We thank you for your interest in TI calculators.
I am copying this message to Herb Foster, Marketing Communications Manager for our calculator group. With Herb's agreement, Texas Instruments grants you permission to copy the materials you describe below for the limited purposes you describe below.
Regards,
Larry Bassuk
Copyright Counsel
972-917-5458

-----Original Message-----
From: Rusen Meylani [mailto:meylani@superonline.com]
Sent: Wednesday, February 20, 2002 5:57 PM
To: copyrightcounsel@list.ti.com - Copyright Legal Counsel
Subject: USAGE OF TI 83+ FACILITIES IN MY SAT II MATH BOOKS
Dear Sir,
I am an educational consultant in Istanbul Turkey and I am working with Turkish students who would like to go to the USA for college education. I am writing SAT II Mathematics books where I make use of TI 83+ facilities, screen shots, etc. heavily. Will you please indicate the copyright issues that I will need while publishing my book?
Thanks very much in advance. I am looking forward to hearing from you soon.
Rusen Meylani.

ACKNOWLEDGMENTS

I would like to thank TEXAS INSTRUMENTS for providing scientists and mathematicians such a powerful hand-held computer, the TI-83 Plus. With this wonderful machine, teachers of mathematics can go beyond horizons without the need to reinvent the wheel all the time. I would also like to thank TEXAS INSTRUMENTS for providing me with a limited copyright to use the graphs that have been produced by the TI-83 Plus graphing calculator throughout this book.

I would like to thank Erdel VIDORI of TEXAS INSTRUMENTS for his suggestions on the organization and title of this book as well as his invaluable efforts in establishing the link between me and TEXAS INSTRUMENTS.

I would like to thank my partner Atakan ARIBURNU for his excellent work on the distribution of this book especially in the United States of America.

I would like to thank Zeynel Abidin ERDEM, the chairman of the Turkish – American Businessmen Association (TABA) for his valuable support and contributions on our projects.

I would like to thank my students, Zekeriyya GEMİCİ, Mehmet Can EGE, Eda UYKAL, Hikmet Can BİLİCİ, and Mehmet Hayri BARAN for their valuable contributions and suggestions on this book.

I would like to thank Pınar ERKORKMAZ for her excellent work on the graphics.

I would like to thank Ahmet ARDUÇ, and my students, the participants of the INTT 121 and INTT 122 classes of Boğaziçi (Bosphorus) University, who helped me in editing and proofreading this book.

I would like to thank Nuran TUNCALI; my teacher, my mentor, and my honorary mother, for inspiring me; teaching me everything I know; for showing me the right path to follow and everything else that she has done and has been doing since the day I had the chance to become her student.

I would like to thank Yorgo İSTEFANOPULOS, Ayşın ERTÜZÜN, Bayram SEVGEN and Zeki ÖZDEMİR, for whatever I know of analytical thinking and mathematics.

Last but not the least, I would like to express my most sincere gratefulness to my mother and father for my being and I dedicate this book to them.

To Mom and Dad …

"Sizi Seviyorum!"

> "Some men see things as they are and say '*Why?*'
> I dream of things that never were and say '*Why not!*' "
> Robert F. KENNEDY

PREFACE

This book is intended to help high school students who are bound to take either or both of the SAT II Mathematics IC and IIC tests. Book 1 is devoted to the usage of the TI – 83 Plus Graphing calculator, particularly in the context of **Algebra, Pre-Calculus and SAT II and IB Mathematics** with over 300 questions carefully designed and fully solved questions. The method proposed in this book has been developed through 5 years' experience; has been proven to work and has created a success story each and every time it was used, having helped hundreds of students who are currently attending the top 50 universities in the USA, that include many Ivy League schools as well.

The main advantage of the approach suggested in this book is that, **YOU CAN SOLVE, ANY TYPE OF EQUATION OR INEQUALITY** with the TI, whether it is algebraic, trigonometric, exponential, logarithmic, polynomial or one that involves absolute values, **WITHOUT NEEDING TO KNOW THE RELATED TOPIC IN DEPTH** and having to perform tedious steps; you can solve all types of equations and inequalities very easily and in a very similar way **JUST NEEDING TO LEARN A FEW VERY EASY TO REMEMBER TECHNIQUES.** But there are still more to what you can do with the TI; find period, frequency, amplitude, offset, axis of wave of a periodic function, find the maxima minima and zeros as well as the domains and ranges of all types of functions; perform any operations on complex numbers, carry out any computation involving sequences and series, perform matrix algebra, solve a system of equations for any number of unknowns and even write small programs to ease your life. More than 20 of the 50 questions in the SAT II Mathematics tests are based on the topics we just have given and this is why this book will definitely raise your SAT II Mathematics scores by at least 200 points.

The topics that are listed above are typically what a student learns at high school algebra classes. This book is intended to fulfill the need for a book specifically designed for an SAT bound college senior who wishes to score perfectly on the SAT II math tests in a very short period of time.

It is very important that a student follows the order in the book step by step, there is nothing more or nothing less than what a student must learn, each and every one of the 300 questions is unique, a wise student must aim at finishing them all, trying to capture the methods suggested.

As a final word for the college bound student; whenever you get tired during the process, remember the words of **Mustafa Kemal ATATÜRK**, one of the greatest leaders of the 20th century, **"Those who are not bound to get any rest, never get tired."**

Please keep in mind, that if you can give yourself to mathematics, mathematics will give you the world. Good luck and be prepared…

Advanced Calculation and Graphing Techniques with the TI – 83 Plus Graphing Calculator

TABLE OF CONTENTS

ACKNOWLEDGMENTS	iii
PREFACE	v
TABLE OF CONTENTS	vii
INTRODUCTION	1
SAT II Mathematics	3
Mathematics Level IC	3
Structure	3
Calculators in the Test	3
Mathematics Level IIC	3
Structure	3
Calculators in the Test	3
Which calculator is allowed and which is not	4
Number of questions per topics covered	4
Similarities and Differences	5
Differences between the tests	5
Geometry	5
Trigonometry	5
Functions	5
Statistics	5
Important Notice on the Scores	6
1. TI BASICS	7
1.1 Turning the TI - 83 Plus On and Off	9
1.2 Resetting Memory	9
1.3 Restoring the Default Settings	9
1.4 Adjusting Screen Contrast	9
1.5 The Math Menu	10
1.6 The Catalog	10
1.7 PEMDAS is observed with the TI	11
1.8 Editing Expressions	12
1.9 The ANS Variable	12
1.10 Accessing a Previous Entry	12
1.11 The Operational Minus Sign and the Number Minus Sign	13
1.12 Number of Floating (Decimal) Points to be Displayed	13
1.13 Storing Values in a Variable	13
1.14 Decimal to Fraction Conversion	14
1.15 Square Roots, Cube Roots, n'th Roots, Fractional Powers	14
1.16 Operations on Complex Numbers	15
1.17 Built in functions in TI that are commonly used in SAT II Math	16
1.18 Additional Functions	17
1.19 Radian and Degree	17
1.20 The numbers " e " (Euler's constant) and " π " (pi)	18

Advanced Calculation and Graphing Techniques with the TI – 83 Plus Graphing Calculator

1.21 Factorial Notation, Permutations and Combinations	18
1.22 Sequences and Series	19
1.24 Matrices and Determinants	20
1.25 Statistics	22
1.26 Simple Programming	24
1.27 Polynomial Root Finder and Simultaneous Equation Solver	25
Polynomial Root Finder	25
Simultaneous Equation Solver	26
2. TI GRAPHING PRELIMINARIES	**27**
2.1 The Y= Editor	29
2.2 Graph Style Icons in the Y= Editor	29
2.3 Graph Viewing Window Settings	30
2.4 Graphing Piecewise Functions	31
2.5 Composition of Functions - Operations and Transformations on Functions	32
2.6 The ZOOM Menu	33
i. Zoom Cursor	33
ii. ZBox	33
iii. Zoom In, Zoom Out	34
iv. ZSquare	35
v. ZStandard	35
vi. ZoomFit	35
2.7 The CALC Menu	35
i. CALC value	36
ii. CALC zero	37
iii. CALC minimum	40
iv. CALC maximum	41
v. CALC intersect	43
2.8 Table	44
2.9 Parametric Graphing	45
2.10 Polar Graphing	47
2.11 Graphing of Conics	47
Circle	47
Ellipse	49
Hyperbola	50
Parabola	52
3. THE METHOD	**55**
Solving Polynomial or Algebraic Equations	57
Solving Absolute Value Equations	57
Solving Exponential and Logarithmic Equations	57
Solving System of Linear Equations	58
Solving Trigonometric Equations	58
Solving Inverse Trigonometric Equations	59
Solving Inequalities	60

Advanced Calculation and Graphing Techniques with the TI – 83 Plus Graphing Calculator

Solving Trigonometric Inequalities	60
Finding Maxima and Minima	62
Finding Domains and Ranges	62
Exploring Evenness and Oddness	62
Graphs of Trigonometric Functions	63
Period	63
Frequency	63
Amplitude	63
Offset	63
Axis of wave equation	63
The Greatest Integer Function	63
TI Usage	63
Parametric Graphing	64
Polar Graphing	64
Limits	64
Existence of Limit	64
Continuity	64
Horizontal and Vertical Asymptotes	64
Zero	65
Hole	65
Vertical asymptote	65
Horizontal asymptote	65
Complex Numbers	65
Permutations and Combinations	65
4. SAMPLE TI PROBLEMS	**67**
4.1 Polynomial Equations	69
4.2 Algebraic Equations	70
4.3 Absolute Value Equations	70
4.4 Exponential and Logarithmic Equations	71
4.5 System of Linear Equations, Matrices and Determinants	71
4.6 Trigonometric Equations	72
4.7 Inverse Trigonometric Equations	73
4.8 Polynomial, Algebraic and Absolute Value Inequalities	73
4.9 Trigonometric Inequalities	74
4.10 Maxima and Minima	74
4.11 Domains and Ranges	74
4.12 Evenness And Oddness	75
4.13 Graphs of Trigonometric Functions	76
4.14 Miscellaneous Graphs	77
4.15 The Greatest Integer Function	78
4.16 Parametric Graphing	78
4.17 Polar Graphing	79
4.18 Limits	79

Advanced Calculation and Graphing Techniques with the TI – 83 Plus Graphing Calculator

4.19 Continuity	80
4.20 Horizontal and Vertical Asymptotes	80
4.21 Complex Numbers	80
4.22 Permutations and Combinations	81
4.23 Miscellaneous Calculations	81
5. SOLUTIONS TO THE SAMPLE TI PROBLEMS	**89**
5.1 Polynomial Equations	91
5.2 Algebraic Equations	98
5.3 Absolute Value Equations	100
5.4 Exponential and Logarithmic Equations	102
5.5 System of Linear Equations, Matrices and Determinants	106
5.6 Trigonometric Equations	108
5.7 Inverse Trigonometric Equations	120
5.8 Polynomial, Algebraic and Absolute Value Inequalities	122
5.9 Trigonometric Inequalities	128
5.10 Maxima and Minima	132
5.11 Domains and Ranges	134
5.12 Evenness And Oddness	138
5.13 Graphs of Trigonometric Functions	146
5.14 Miscellaneous Graphs	151
5.15 The Greatest Integer Function	155
5.16 Parametric Graphing	158
5.17 Polar Graphing	163
5.18 Limits	165
5.19 Continuity	168
5.20 Horizontal and Vertical Asymptotes	170
5.21 Complex Numbers	174
5.22 Permutations and Combinations	177
5.23 Miscellaneous Calculations	178
ANSWERS	208
Sample SAT II Mathematics Level IC Model Test	217
Answer Sheet	233
Answer Key	235
Score Conversion Table	236
Sample SAT II Mathematics Level IIC Model Test	237
Answer Sheet	253
Answer Key	255
Score Conversion Table	256
INDEX	257

INTRODUCTION

— Advanced Calculation and Graphing Techniques with the TI – 83 Plus Graphing Calculator —

SAT II Mathematics

Mathematics Level IC and Mathematics Level IIC are the two subject tests that the College Board offers in SAT II Mathematics. Both tests require at least a scientific, preferably a graphing, calculator. Each test is one hour long.

Mathematics Level IC

Structure

A Mathematics Level IC test is made of 50 multiple choice questions from the following topics:

- Algebra and algebraic functions
- Geometry (plane Euclidean, coordinate and three-dimensional)
- Elementary statistics and probability, data interpretation, counting problems, including measures of mean, median and mode (central tendency.)
- Miscellaneous questions of logic, elementary number theory, arithmetic and geometric sequences.

Calculators in the Test

Approximately 60 percent of the questions in the test should be solved without the use of the calculator. For the remaining 40 percent, the calculator will be useful if not necessary.

Mathematics Level IIC

Structure

A Mathematics Level IIC test also is made of 50 multiple choice questions. The topics included are as follows:

- Algebra
- Geometry (coordinate geometry and three-dimensional geometry)
- Trigonometry
- Functions
- Statistics, probability, permutations, and combinations
- Miscellaneous questions of logic and proof, elementary number theory, limits and sequences

Calculators in the Test

In Math IIC, 40 percent of the questions should be solved the without the use of the calculator. In the remaining 60 percent, the calculator will be useful if not necessary.

Advanced Calculation and Graphing Techniques with the TI – 83 Plus Graphing Calculator

Which calculator is allowed and which is not:

The simplest reference to this question is this: No device with a QWERTY keyboard is allowed. Besides that any hand held organizers, mini or pocket computers, laptops, pen input devices or writing pads, devices making sounds (Such as "talking" computers) and devices requiring electricity from an outlet will not be allowed. It would be the wisest to stick with TI 83 or TI 89. Both of these calculators are easy to use and are the choices of millions of students around the world who take SAT exams and also university students in their math courses. It is very important to be familiar with the calculator that you're going to use in the test. You will lose valuable time if you try to figure it out during the test time.

Be sure to learn to solve each and every question in this book. They are carefully chosen to give you handiness and speed with your calculator. You will probably gain an extra 150 to 200 points in a very short period of time.

IMPORTANT: Always take the exam with fresh batteries. Bring fresh batteries and a backup calculator to the test center. You may not share calculators. You certainly will not be provided with a backup calculator or batteries. No one can or will assist you in the case of a calculator malfunction. In such case, you have the option of notifying the supervisor to cancel your scores for that test. Therefore, always be prepared for the worst case scenario (Don't forget Murphy's Rules.)

Number of questions per topics covered

The following chart shows the approximate number of questions per topic for both tests.

Topics	Approximate Number of Questions	
	Level IC	Level IIC
Algebra	15	9
Plane Euclidean Geometry	10	0
Coordinate Geometry	6	6
Three-dimensional Geometry	3	4
Trigonometry	4	10
Functions	6	12
Statistics	3	3
Miscellaneous	3	6

Similarities and Differences

Some topics are covered in both tests, such as elementary algebra, three-dimensional geometry, coordinate geometry, statistics and basic trigonometry. But the tests differ greatly in the following areas.

Differences between the tests

Although some questions may be appropriate for both tests, the emphasis for Level IIC is on more advanced content. The tests differ significantly in the following areas:

Geometry

Euclidian geometry makes up the significant portion of the geometry questions in the Math IC test. Though in IIC, questions are of the topics of coordinate geometry, transformations, and three-dimensional geometry and there are no direct questions of Euclidian geometry.

Trigonometry

The trigonometry questions on Level IC are primarily limited to right triangle trigonometry and the fundamental relationships among the trigonometric ratios. Level IIC places more emphasis on the properties and graphs of the trigonometric functions, the inverse trigonometric functions, trigonometric equations and identities, and the laws of sines and cosines. The trigonometry questions in IIC exam are primarily on graphs and properties of the trigonometric functions, trigonometric equations, trigonometric identities, the inverse trigonometric functions, laws of sines and cosines. On the other hand, the trigonometry in IC is limited to basic trigonometric ratios and right triangle trigonometry.

Functions

Functions in IC are mostly algebraic, while there are more advanced functions (exponential and logarithmic) in IIC.

Statistics

Probability, mean median, mode counting, and data interpretation are included in both exams. In addition, IIC requires permutations, combinations, and standard deviation.

Advanced Calculation and Graphing Techniques with the TI – 83 Plus Graphing Calculator

In all SAT Math exams, you must choose the best answer which is not necessarily the exact answer. The decision of whether or not to use a calculator on a particular question is your choice. In some questions the use of a calculator is necessary and in some it is redundant or time consuming. Generally, the angle mode in IC is degree. Be sure to set your calculator in degree mode by pressing "Mode" and then selecting "Degree." However, in IIC you must decide when to use the "Degree" mode or the "Radian" mode. There are figures in some questions intended to provide useful information for solving the question. They are accurate unless the question states that the figure is not drawn to scale. In other words, figures are correct unless otherwise specified. All figures lie in a plane unless otherwise indicated. The figures must NOT be assumed to be three-dimensional unless they are indicated to be. The domain of any function is assumed to be set of all real numbers x for which $f(x)$ is a real number, unless otherwise specified.

Important Notice on the Scores

In SAT IC questions the topics covered are relatively less than those covered in the IIC test. However, the questions in the IC exam are more tricky compared to the ones in IIC. This is why if students want to score 800 in the IC test, they have to answer all the 50 questions correctly. But in the IIC test, 43 correct answers (the rest must be omitted) are sufficient to get the full score of 800.

Scaled Score	Raw Score in IC	Raw Score in IIC
800	50	43
750	45	38
700	38	33
650	33	28
600	29	22
550	24	16
500	19	10
450	13	3
400	7	0
350	1	-3

CHAPTER 1.
TI BASICS

Advanced Calculation and Graphing Techniques with the TI – 83 Plus Graphing Calculator

1.1 Turning the TI - 83 Plus On and Off

In order to turn on the TI – 83 Plus, press the [ON] key.

In order to turn off the TI- 83 Plus press the [2nd] key followed by the [ON] key.

1.2 Resetting Memory

Resetting the memory will restore the initial factory settings that are the default settings of the calculator as well as clear the memory.

1.3 Restoring the Default Settings

This may be necessary when the default calculator settings have to be retrieved in a single step after changing mode settings, screen settings, etc.

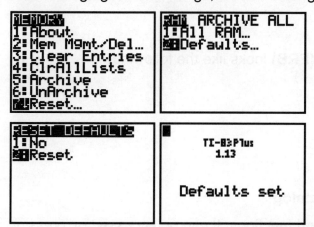

1.4 Adjusting Screen Contrast

When the calculator is reset, it will retain its factory settings, which sometimes cause the screen to have non-ideal contrast adjustment. In order to achieve proper contrast adjustment:

Press the [2nd] key followed by the [▲] or the [▼] keys to darken or lighten the screen.

1.5 The Math Menu

Most of the built in commands that will be used in the SAT II Math context, that don't exist as separate keys or key combinations, can be found in the **MATH** menu, followed by the left, right arrows and up, down arrows.

The **MATH** main screen can be accessed by pressing the  key and looks like the following:

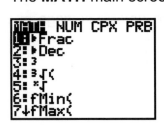

The numerical calculations **(NUM)** screen looks like the following:

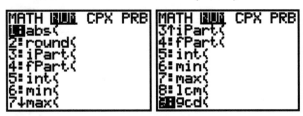

The complex calculations **(CPX)** screen looks like the following:

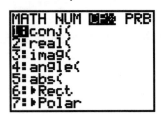

The screen for the calculations involving probability **(PRB)** looks like the following:

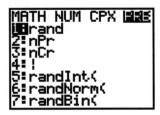

1.6 The Catalog

The catalog contains all of the built in expressions and functions. It can be accessed by pressing the 2nd key followed by the 0 key that will serve as the **CATALOG** command. **Pressing any key that contains a capital letter on the right hand corner (without the ALPHA key) will take you to the beginning of the portion of the catalog whose initial is that letter.** For example after the catalog is opened, pressing the **SIN** key will take you to the beginning of the commands staring with the letter **E**. After this point the arrows followed by **ENTER** may be used to select the appropriate command. **CATALOG** can be accessed when the exact location of a certain expression or command is forgotten.

1.7 PEMDAS is observed with the TI

PEMDAS is the correct mathematical order of operations, which is observed with the TI. Therefore you do not need to open redundant parentheses and complicate simple mathematical expressions. Any correct mathematical expression will be calculated correctly subject to the **PEMDAS** rule.

Parentheses
Exponents
Multiplication   The order of operations
Division
Addition
Subtraction

$10 + 2 \cdot 3^2 - 9^{(2+1)} / 18 \times 2$

must be input as:

```
10+2*3²-9^(2+1)/
18*2
              -53
```

On the other hand, $4/3\pi$ will be assumed by TI as $\frac{4}{3}\pi$ and not as $\frac{4}{3\pi}$. When $\frac{4}{3\pi}$ is meant it must be input as $4/(3\pi)$; in such case the use of a parenthesis will be absolutely necessary.

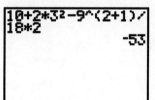

Similarly $4/3\sqrt{2}$ will be assumed by TI as $\frac{4}{3}\sqrt{2}$ and not as $\frac{4}{3\sqrt{2}}$. When $\frac{4}{3\sqrt{2}}$ is meant it must be input as $4/(3\sqrt{2})$; in such a case the use of a parenthesis is necessary.

Advanced Calculation and Graphing Techniques with the TI – 83 Plus Graphing Calculator

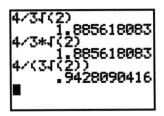

1.8 Editing Expressions

Any expression can be edited using a combination of the following keys:

The **CLEAR** key when pressed once, will clear a line; when pressed twice, will clear the whole screen. The **CLEAR** key may also be used to leave a current screen and go back to the home (initial) screen. In any case of error or when **CLEAR** key does not work, the 2nd key followed by the **MODE** key will serve as the **QUIT** command. The **DEL** key will delete exactly one character where the cursor is positioned each time it is pressed. The 2nd key followed by the **DEL** key will act as the **INS** (insert key) and it may be used to insert an expression exactly before the position of the cursor. In order to position the cursor, a combination of the left, right, up, down arrows can be used.

1.9 The ANS Variable

The last answer after the enter key is pressed is stored in the variable **ANS**:

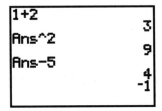

The **ANS** variable is updated whenever the enter key is pressed. If there is an operation associated with the **ANS** variable or if there is an expression involving the **ANS** variable prior to pressing the enter key, the output will be changed each time the enter key is pressed. For example if the last expression entered is **ANS-5** and the last answer is 4, pressing the enter key once again will produce –1, which is the last answer decreased by 5.

1.10 Accessing a Previous Entry

In order to access previous entries that you have made, for each retrieval, press the key followed by the key. This will retrieve the previous entries since the last reset in a cyclic fashion depending on the memory **(RAM)** of your calculator. Please note that the content if the **ANS** variable is modified each time the enter key is pressed.

Advanced Calculation and Graphing Techniques with the TI – 83 Plus Graphing Calculator

1.11 The Operational Minus Sign and the Number Minus Sign

The TI has two minus keys, which are often confused. The one on the left hand side above is the number minus sign and it will be used while writing $(-3+4)^2$. The one on the right hand side is the operational minus sign and it must be used when writing $(3-4)^2$. These two keys mixed up may lead to syntax errors or which is worse, calculation errors and incorrect answers.

1.12 Number of Floating (Decimal) Points to be Displayed

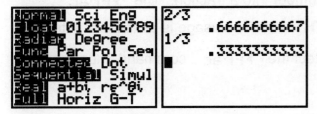

Please note that the displayed answer is not actually changed by the calculator by a rounding off operation. The displayed number of decimal points is 3 but the actual results that are stored in memory still have the maximum number of decimal points. The display is changed only, the results are not changed at all.

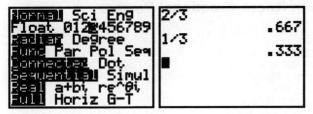

1.13 Storing Values in a Variable

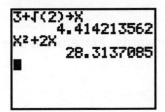

In the above calculation, $3+\sqrt{2}$ is stored in X and X^2+2X is evaluated. In order to do this operation the following key combination must be used:

$3+\sqrt{2}$ STO▸ X

You can also store different numbers in different variables and work accordingly:

For Example:

X=3.7824 and Y=2π-3 and you are required to calculate the following:

$$\frac{X^2 - XY}{X + Y^3}$$

In this case you should do the following in TI:

1.14 Decimal to Fraction Conversion

Sometimes it may be necessary to interpret a result as a fraction rather than a decimal. TI has a powerful tool that allows us to convert a decimal number into a fraction. The ▶Frac command can be found as the first entry in the **MATH** menu. The usage is as follows: After the result is displayed, press the (MATH) key to select the ▶Frac command, then press the enter key.

1.15 Square Roots, Cube Roots, n'th Roots, Fractional Powers

$\sqrt{7+5}$ must be input as:

Please note that there is no operation associated with the parenthesis afterwards, therefore it does not need to be closed.

$\sqrt[3]{9^2}$ can be input in one of the following ways:

Please be careful with how you enter $9^{\frac{2}{3}}$. 9^2/3 would mean $(9^2)/3$ and not $9^{\frac{2}{3}}$ or $9^{2/3}$. $9^{\frac{2}{3}}$ must be entered as 9^(2/3 or 9^(2/3). Please also note that, the $\sqrt[x]{}$ operation does not automatically open a parenthesis, therefore you should open it manually.

Advanced Calculation and Graphing Techniques with the TI – 83 Plus Graphing Calculator

1.16 Operations on Complex Numbers

On the first hand the calculator mode must be set to **a+bi** as follows:

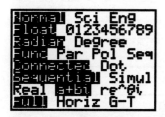

i is the imaginary number that has the following properties:

$i^2 = -1$ and $i = \sqrt{-1}$

$\dfrac{(3-2i)}{(1+i)^3}$ must be input as:

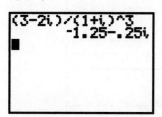

Other operations that are commonly needed in the SAT II Math context are as follows:

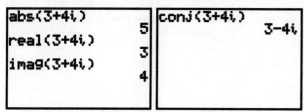

The required functions can be found at the **MATH CPX** menu:

Advanced Calculation and Graphing Techniques with the TI – 83 Plus Graphing Calculator

Unauthorized copying or reuse of any part of this page is illegal.

1.17 Built in functions in TI that are commonly used in SAT II Math

FUNCTION	DESCRIPTION	ABBREVIATION
sin(	Sine function	
cos(	Cosine function	
tan(	Tangent function	
$\sin^{-1}($	Arcsine or sine inverse function	Arcsin
$\cos^{-1}($	Arccosine or cosine inverse function	Arccos
$\tan^{-1}($	Arctangent or tangent inverse function	Arctan
10^(		
e^(		
log(	Logarithm base 10	
ln(	Logarithm base e	
$\sqrt{($	Square root	
abs(	Absolute value	
int(	Greatest integer function	

As an example $\dfrac{1+3\ln 2}{4-3\cdot e^5}$ is input as follows:

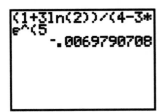

The **abs(** and **int(** functions can be found using the **MATH NUM** menu:

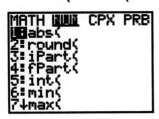

Advanced Calculation and Graphing Techniques with the TI – 83 Plus Graphing Calculator

1.18 Additional Functions

WHEN YOU NEED	USE THE FOLLOWING DEFINITION	WARNING
$\sec(x)$	$1/\cos(x)$	
$\csc(x)$	$1/\sin(x)$	
$\cot(x)$	$\cos(x)/\sin(x)$	Although mathematically correct, still do NOT use $1/\tan(x)$ for $\cot(x)$ because when $\tan(x)$ is undefined, TI will interpret $1/\tan(x)$ as undefined, too, which is not correct.
$\sec^{-1}(x)$	$\cos^{-1}(1/x)$	Do NOT use $1/\cos^{-1}(x)$, mathematically incorrect.
$\csc^{-1}(x)$	$\sin^{-1}(1/x)$	Do NOT use $1/\sin^{-1}(x)$, mathematically incorrect.
$\cot^{-1}(x)$	$\pi/2 - \tan^{-1}(x)$ if in radians $90° - \tan^{-1}(x)$ if in degrees	Do NOT use $1/\tan^{-1}(x)$, mathematically incorrect.
$\log_a x$	$\log(x)/\log(a)$	Either of these definitions can be used but please be careful with closing the parentheses that are automatically opened when **LOG** or **LN** are pressed.
	$\ln(x)/\ln(a)$	

As an Example: $\log_2 3$ can be input in one of the following ways:

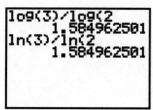

1.19 Radian and Degree

Radian mode interprets angle values as radians and answers displayed are also in radians.

Advanced Calculation and Graphing Techniques with the TI – 83 Plus Graphing Calculator

Degree mode interprets angle values as degrees and answers displayed are also in degrees.

1.20 The numbers " e " (Euler's constant) and " π " (pi)

The number e can be input by the following key combination: [2nd] [÷]

The number π can be input by the following key combination: [2nd] [^]

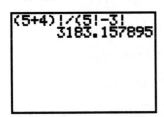

1.21 Factorial Notation, Permutations and Combinations

n! = The product of all consecutive integers starting with 1 up to and including n that is:

n! = n . (n-1) . (n-2) . (n-3) . . . 3 . 2 . 1

$\dfrac{(5+4)!}{5!-3!}$ can be input as follows:

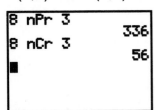

P(n,r): Number of permutations of r elements chosen from n elements where $r \leq n$

$P(n,r) = \dfrac{n!}{(n-r)!}$

C(n,r): Number of combinations of r elements chosen from n elements where $r \leq n$

$C(n,r) = \dfrac{n!}{(n-r)! \cdot r!}$

P(8,3) and C(8,3) can be input respectively as follows:

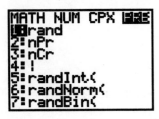

The required functions can be found at the **MATH PRB** menu:

1.22 Sequences and Series

The TI has the built in feature of **seq(** command in the **LIST OPS** menu that can be accessed through the (2nd) key followed by the (STAT) key, for analyzing sequences.

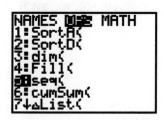

The **seq(** command will be used to generate the terms of a sequence with the following method of usage:

seq(the formula for the sequence, the variable of the sequence, the starting value for the variable, the ending value for the variable, the increment)

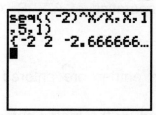

In the above example the following inputs were used:

the formula for the sequence: $\dfrac{(-2)^x}{x}$

the variable of the sequence: x

the starting value for the variable: 1

the ending value for the variable: 5

the increment: 1

The sum of the terms of the sequence i.e. the series can be found using the **sum(** command in the **LIST MATH** menu.

Please remember that for arithmetic sequences the formula for the sequence must be input as **a+(x-1) · d** where **a** is the first term of the sequence and **d** is the common difference between a term and the next.

For example for the sequence 8, 11, 14, 17, 20,… the formula for the sequence will be: **8+(x-1) · 3**

Please also remember that for geometric sequences the formula for the sequence must be input as **a · r**[(x-1)] where **a** is the first term of the sequence and **r** is the common ratio between a term and the next.

For example for the sequence 5, 10, 20, 40, 80,... the formula for the sequence will be: **5 · 2**[(x-1)]

1.24 Matrices and Determinants

In order to solve a system of linear equations, the coefficients of the linear system are entered in two separate matrices A and B followed by the simple algebraic operation of $A^{-1} \cdot B$. All arithmetic operations can be performed similarly.

The matrix menu is accessed by pressing the $\boxed{x^{-1}}$ key. The matrix entries are entered through the **EDIT** sub menu where the dimensions of the matrix followed by the matrix entries must be input. The matrices that are entered will be used through the **NAMES** sub menu later on. In the following example A is a 2 by 2 matrix and B is a 2 by 1 matrix.

Example:

$$\left. \begin{array}{l} x + 3y = 7 \\ 12x - 2y = 8 \end{array} \right\} \frac{x}{y} = ?$$

Solution:

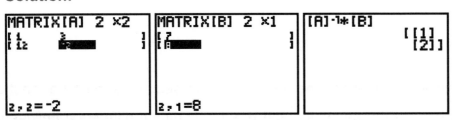

Answer: 1/ 2

Advanced Calculation and Graphing Techniques with the TI – 83 Plus Graphing Calculator

Example:

$$\left.\begin{array}{l} x + y + z = 6 \\ 2x - y + 3z = 9 \\ 3x + y - 4z = -7 \end{array}\right\} x^2+y^2+z^2=?$$

Solution:

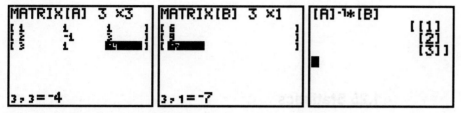

Answer: $1^2+2^2+3^2 = 1+4+9 = 14$

Example:

If $A = \begin{bmatrix} 2 & 3 \\ -1 & 5 \end{bmatrix}$, $B = \begin{bmatrix} 0 & -2 \\ -3 & 5 \end{bmatrix}$, Find

(i) $A + B$; (ii) $3A - 2B$; (iii) AB; (iv) BA

Solution:

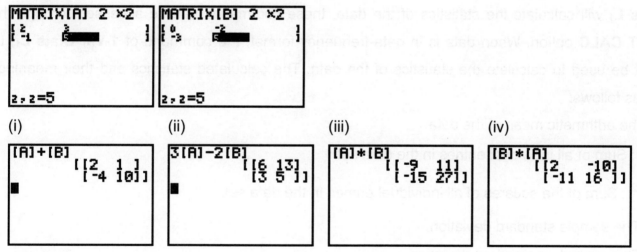

Example:

Find the determinant and inverse of the matrix $\begin{bmatrix} 1 & 3 & -1 \\ -2 & 4 & 1 \\ 0 & 0 & 2 \end{bmatrix}$

Solution:

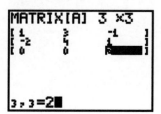

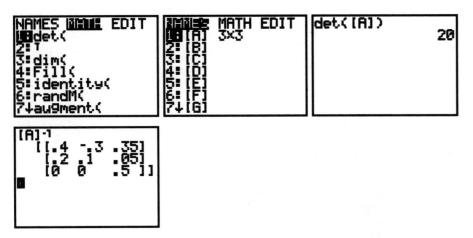

1.25 Statistics

Before performing any statistical calculations, it may be essential to clear any previous list entries. **MEMory ClrAllLists** option will accomplish this task. The data may be given in two ways, in raw format or in data-frequency format. When data is given in raw format, all data is entered in the list named L_1 using the **STAT EDIT** option. When data is given in data-frequency format, the data is entered in the list named L_1 and the individual frequencies are entered in the list named L_2, again using the **STAT EDIT** option. When data is in raw format, the command of **1-Var Stats** or **1-Var Stats L_1** will calculate the statistics of the data, these commands can be accessed through the **STAT CALC** option. When data is in data-frequency format, the command of **1-Var Stats L_1, L_2** must be used to calculate the statistics of the data. The calculated statistics and their meanings are as follows:

$\bar{x}$: The arithmetic mean of the data.

$\sum x$: Sum of all individual entries in the data set.

$\sum x^2$: Sum of the squares of all individual entries in the data set.

Sx: The sample standard deviation.

σx: The population standard deviation.

n: number of data

minX: Minimum entry in the data set.

Q_1: Lower Quartile

Med: Median

Q_3: Upper Quartile

MaxX: Maximum entry in the data set.

Example:

Find the statistics of the following data: 1, 3, 5, 6, 3, 6, 6.

Advanced Calculation and Graphing Techniques with the TI – 83 Plus Graphing Calculator

Solution:

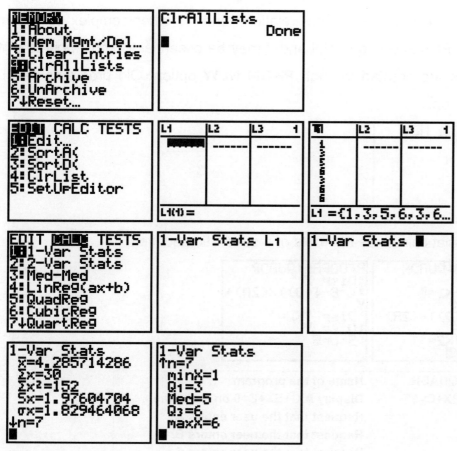

Example:

Find the statistics of the following data:

Data	1	3	5	6
Frequency	1	2	1	3

Solution:

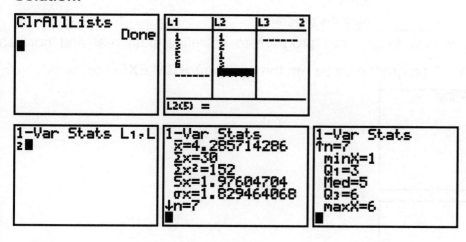

1.26 Simple Programming

Writing simple programs involving only input-output relations and no other complex conditional statements is easy with TI- 83 Plus and it is useful and it may be essential to know how to write one. First of all new programs are created through **PRGM NEW** option. Old programs can be edited through **PRGM EDIT** option.

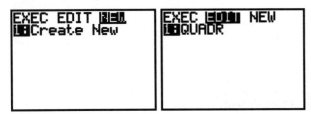

Following is a simple program that calculates the roots of a quadratic equation.

PROGRAM: QUADR	Name of the program
:Disp "AX²+BX+C=0"	Display AX²+BX+C=0 on the screen.
:Prompt A	Request that the user enters A.
:Prompt B	Request that the user enters B.
:Prompt C	Request that the user enters C.
:B²-4AC→D	Calculate B²-4AC and store in D
:(-B+√(D))/(2A)→X	Calculate (-B+√(D))/(2A) and store in X.
:Disp "X1="	Write X1= on the screen.
:Disp X	Write X.
:(-B-√(D))/(2A)→Y	Calculate (-B-√(D))/(2A) and store in Y.
:Disp "X2="	Write X2= on the screen.
:Disp Y	Write Y.
:Stop	Stop the program.

Following is a demonstration of how to use this program to calculate both real and non-real solutions of a quadratic equation. A program may be run through the **PRGM EXEC** option.

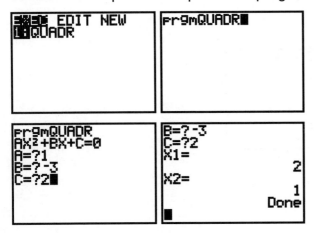

Advanced Calculation and Graphing Techniques with the TI – 83 Plus Graphing Calculator

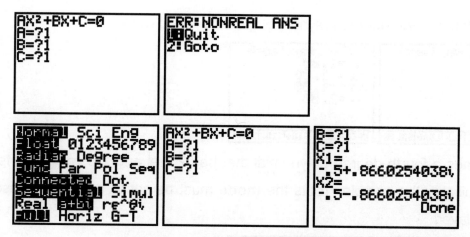

1.27 Polynomial Root Finder and Simultaneous Equation Solver

In the TI-83 Plus there is the **APPS** key that enables access to several useful applications. These applications are either built in or downloadable from the TI site **education.ti.com**. A few of these applications are especially designed to make the lives of a college bound high school student easier. One of these applications is named as **PolySmlt** that makes it possible to find the real and complex roots of a polynomial function and to find the solution of a system of simultaneous equations by a special application accessible through the combination of **APPS PolySmlt**.

Polynomial Root Finder

The first sub-menu in the **APPS PolySmlt** menu is the **Polynomial Root Finder**.

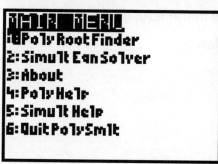

While using this application the degree of the polynomial must be entered first. The next step is to enter the coefficients of the terms in the polynomial. If the degree of the polynomial is **n** then **n+1** coefficients must be entered.

The following example illustrates a fourth degree polynomial that has 4 real roots.

The following example illustrates a fourth degree polynomial that has 2 real and 2 complex roots. Please note that in order to display the complex roots the **mode** must be set to **a+bi**. Otherwise complex roots will not be displayed.

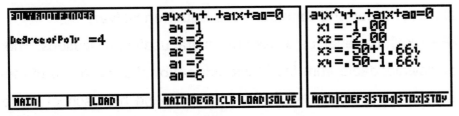

Simultaneous Equation Solver

The second sub-menu in the **APPS PolySmlt** menu is the **Simultaneous Equation Solver**.

While using this application the number of unknowns and the number of equations must be entered first. The next step is to enter the coefficients of the terms in the equations.

The following example illustrates the solution of the following system:

x+2y+4z=3

2x+3y+z=5

x+4y+2z=1

The **Simultaneous Equation Solver** gives the solution as x=3.5; y=-0.75; and z=0.25.

CHAPTER 2.
TI GRAPHING
PRELIMINARIES

Advanced Calculation and Graphing Techniques with the TI – 83 Plus Graphing Calculator

2.1 The Y= Editor

The graphs of functions etc. can be input using the Y= editor. The Y= editor can be accessed by pressing the [Y=] key.

Mode setting	The variable that appears when the [x,T,θ,n] key is pressed
Function	X
Parametric	T
Polar	θ

2.2 Graph Style Icons in the Y= Editor

The following table describes the graph styles available for function graphing. Use the styles to visually differentiate functions to be graphed together. For example, you can set Y_1 as a solid line, Y_2 as a dotted line, and Y_3 as a thick line as follows:

ICON STYLE	NAME	DESCRIPTION
╲	Line	A solid line connects plotted points; this is the default in Connected mode.
╲	Thick	A thick solid line connects plotted points.
▜	Above	Shading covers the area above the graph. This is the style that must be used for plotting $y>f(x)$ or $y \geq f(x)$ type of graphs.
▙	Below	Shading covers the area below the graph. This is the style that must be used for plotting $y<f(x)$ or $y \leq f(x)$ type of graphs.
⋰	Dot	A small dot represents each plotted point; this is the default in Dot mode. This is the style that must be used for plotting functions involving jump discontinuities such as the greatest integer function, i.e. int(x) so that false vertical lines will be eliminated while plotting the graphs.

Advanced Calculation and Graphing Techniques with the TI – 83 Plus Graphing Calculator

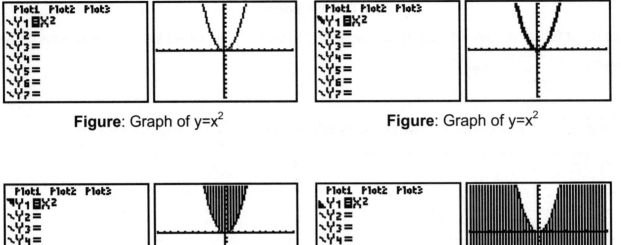

Figure: Graph of $y=x^2$ **Figure**: Graph of $y=x^2$

Figure: Graph of $y>x^2$ **Figure**: Graph of $y<x^2$

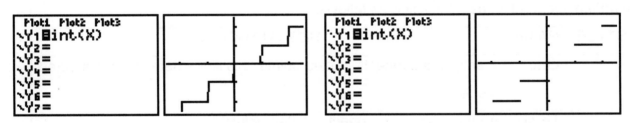

The style icons can be changed by using the 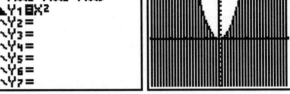 key while in Y= editor in order to move on to the style icon and then by pressing the ENTER key repeatedly until the desired style is. displayed

2.3 Graph Viewing Window Settings

The viewing window is the portion of the coordinate plane defined by **Xmin, Xmax, Ymin,** and **Ymax. Xscl** (X scale) defines the distance between tick marks on the x-axis. **Yscl** (Y scale) defines the distance between tick marks on the y-axis. To turn off tick marks, set **Xscl=0** and **Yscl=0.**

Advanced Calculation and Graphing Techniques with the TI – 83 Plus Graphing Calculator

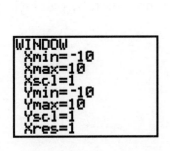
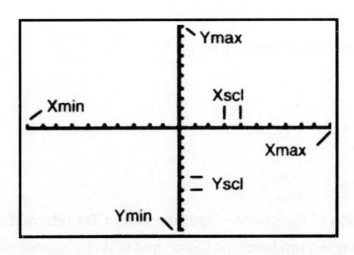

Viewing window settings can be accessed by pressing the (WINDOW) key.

2.4 Graphing Piecewise Functions

Piecewise Functions are those that are not defined by a single unique formula. TI allows the user to enter and plot such functions, too. Let us consider the following example:

$$f(x) = \begin{cases} 2 & x \leq -2 \\ x^2 & -2 < x < 1 \\ -x+3 & x \geq 1 \end{cases}$$

The above function must be entered and will be graphed as follows:

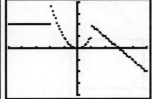

Please note that the **Graph Style** is changed to **Dot** (∴) in order to avoid **false vertical lines** that are demonstrated in the following screen capture:

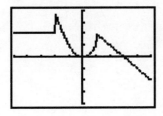

The <,>, ≤, ≥ symbols can be accessed through the key combination by displaying the following menu:

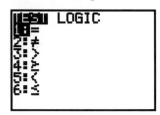

2.5 Composition of Functions - Operations and Transformations on Functions

Let us say that we are given two functions $f(x)=x^2$ and $g(x)=1-x$ and we would like to plot $(f \circ g)(x)=f(g(x))$ and $(g \circ f)(x)=g(f(x))$.

 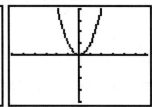  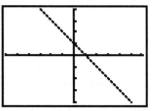

The following screen captures correspond to $(f \circ g)(x)=(1-x)^2$.

  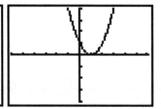

The following screen captures correspond to $(g \circ f)(x)=1-x^2$.

 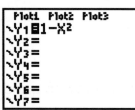 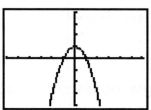

The following screen captures correspond to $(f+g)(x)=f(x)+g(x)=x^2+1-x$.

 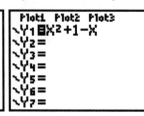 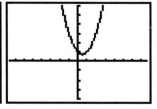

Advanced Calculation and Graphing Techniques with the TI – 83 Plus Graphing Calculator

Unauthorized copying or reuse of any part of this page is illegal.

The following screen captures correspond to $f(x-1)+1=(x-1)^2+1$.

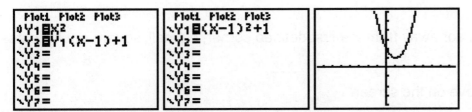

In the preceding examples, Y_1, Y_2, Y_3, etc. have been entered by pressing the (VARS) key and through the **Y-VARS FUNCTION** menu.

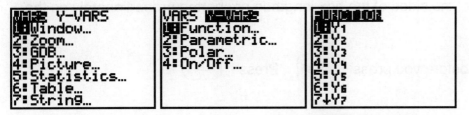

2.6 The ZOOM Menu

The **ZOOM** menu can be accessed by pressing the (ZOOM) key.

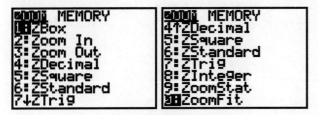

Commonly used **ZOOM** functions in the SAT II Math context are as follows:

ZBox	Draws a box to define the viewing window.
Zoom In	Magnifies the graph around the cursor.
Zoom Out	Views more of a graph around the cursor.
ZSquare	Sets equal-size pixels on the X and Y axes.
ZStandard	Sets the standard window variables.
ZoomFit	Fits Ymin and Ymax between Xmin and Xmax.

i. Zoom Cursor

When the following are selected: **1:ZBox, 2:Zoom In, 3:Zoom Out;** the cursor on the graph becomes the **zoom cursor** (+), a smaller version of the free-moving cursor.

ii. ZBox

To define a new viewing window using **ZBox,** follow these steps.

1. Select **1:ZBox** from the **ZOOM** menu. The zoom cursor is displayed at the center of the screen.

2. Move the zoom cursor to any spot you want to define as a corner of the box, and then press .

When you move the cursor away from the first defined corner, a small, square dot indicates the spot.

3. Press arrow keys to move on the screen:

As you move the cursor, the sides of the box lengthen or shorten proportionately on the screen.

Note: To cancel **ZBox** before you press [ENTER] Press .

4. When you have defined the box, press [ENTER] to re-plot the graph.

The following example demonstrates the usage of the **ZBox** facility.

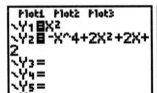 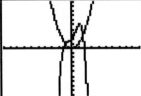

  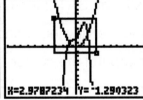 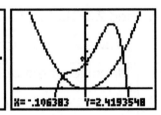

iii. Zoom In, Zoom Out

Zoom In magnifies the part of the graph that surrounds the cursor location.

Zoom Out displays a greater portion of the graph, centered on the cursor location.

To **zoom in** on a graph, follow these steps.

　　Select **2:Zoom In** from the **ZOOM** menu. The zoom cursor is displayed.

　　Move the zoom cursor to the point that is to be the center of the new viewing window.

　　Press .

To **zoom out** on a graph, follow these steps.

　　Select **3:Zoom Out** from the **ZOOM** menu. The zoom cursor is displayed.

　　Move the zoom cursor to the point that is to be the center of the new viewing window.

Advanced Calculation and Graphing Techniques with the TI – 83 Plus Graphing Calculator

Press [ENTER].

Note: To cancel Zoom In or Zoom Out, press [CLEAR].

iv. ZSquare

ZSquare re-plots the functions immediately. It redefines the viewing window based on the current values of the window variables. It adjusts in only one direction so that **ΔX=ΔY,** which makes the graph of a circle look like a circle. **Xscl** and **Yscl** remain unchanged. The midpoint of the current graph (not the intersection of the axes) becomes the midpoint of the new graph.

v. ZStandard

ZStandard re-plots the functions immediately. It updates the window variables to the standard values shown below.

Xmin = -10   Ymin = -10
Xmax = 10   Ymax = 10
Xscl = 1   Yscl = 1

vi. ZoomFit

ZoomFit re-plots the functions immediately. **ZoomFit** recalculates **Ymin** and **Ymax** to include the minimum and maximum Y values of the selected functions between the current **Xmin** and **Xmax**. **Xmin** and **Xmax** are not changed.

2.7 The CALC Menu

The **CALC** menu can be accessed by pressing the  key followed by the key.

The **CALC** menu items that will be heavily used in the context of SAT II Math are **value**, **zero**, **minimum**, **maximum** and **intersect**.

```
CALCULATE
1:value
2:zero
3:minimum
4:maximum
5:intersect
6:dy/dx
7:∫f(x)dx
```

Suppose we would like to find the zeros, the local maximum and the local minimum point of the following function. Suppose we would also like to find the intersection point(s) of this function with another function that will be given.

Firstly the function $y = x^3 - x^2 - 2x + \frac{1}{2}$ will be plotted and zoomed in once to get a clear viewing of the zeros, maximum and minimum point.

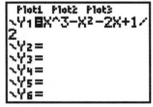   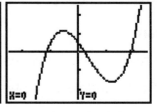

i. CALC value

The **value** facility will calculate the y-coordinate of a point whose x-coordinate that must be within the viewing window is to be entered by the user. The x-coordinate can be any value, positive, or negative within the viewing window and multiple y coordinates can be calculated one after another by selecting **CALC value** only once.

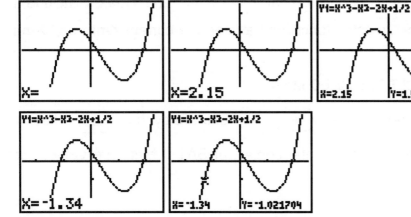

The x-coordinate to be entered can also be any correct mathematical expression that is within the viewing window. The expression will be calculated and converted to a decimal number that will be the corresponding x-coordinate and then the y-coordinate will be calculated similarly.

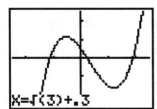 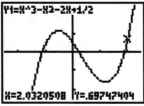

ii. CALC zero

The zeros of a function are the points, whose y-coordinates are zero. The x coordinates of such points are also named as the "roots" of this function. The **CALC zero** facility will enable the user to find such points that lie within the viewing window.

When **CALC zero** is selected, the user will be prompted to select the **Left Bound**, and the **Right Bound** of the zero that is to be found. Since zero is a point where the graph of the function "cuts" the x-axis, the graph has to have a sign change in the neighborhood of the zero. That is, in the neighborhood of the zero, it has to go **from negative (below the x-axis) to positive (above the x-axis) demonstrating an increasing function** or **from positive (above the x-axis) to negative (below the x-axis) demonstrating a decreasing function**.

So the graph will look like either one of the following in the neighborhood of the zero:

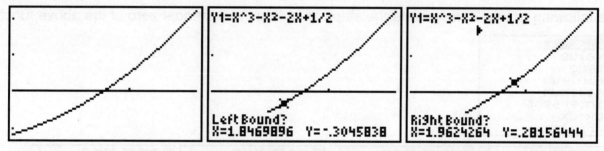

Figure Graph goes from negative (below the x-axis) to positive (above the x-axis):

Increasing behavior in the neighborhood of the zero.

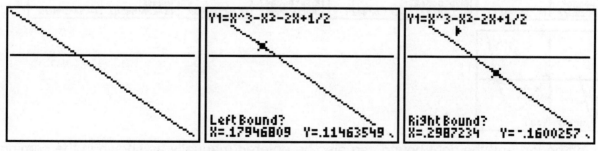

Figure Graph goes from positive (above the x-axis) to negative (below the x-axis):

Decreasing behavior in the neighborhood of the zero.

If the **graph** cuts the x-axis in an **increasing** fashion, the **left bound** must be given **from below** the x-axis and the **right bound** must be given **from above** the x-axis as is in the upper set of the preceding graphs.

If the **graph** cuts the x-axis in a **decreasing** fashion, the **left bound** must be given **from above** the x-axis and the **right bound** must be given **from below** the x-axis as is in the lower set of the preceding graphs.

After the **left bound** and the **right bound** are entered correctly, there will be the **guess**ing step which may be passed quickly by pressing the enter key once more.

If the left bound and the right bound are not entered correctly, one of the following error messages will appear on the screen.

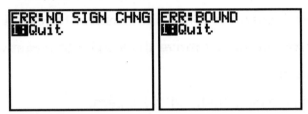

The following graphs demonstrate the steps of finding the rightmost zero of the above function.

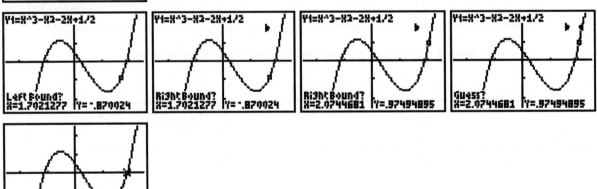

The left bound and the right bound can also be entered manually as shown in the coming graphs. In this case, the left bound will be the x-coordinate of a point to the left of the zero and the right bound will be the x-coordinate of a point to the right of the zero. This option is especially useful when the cursor does not appear on the screen or when the graph increases or decreases so steeply that the locations of the left bound and the right bound cannot be very easily seen.

Advanced Calculation and Graphing Techniques with the TI – 83 Plus Graphing Calculator

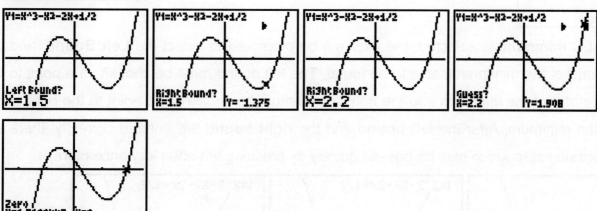

The other zeros of the above function are as follows:

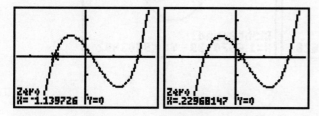

If the graph of the function is tangent to the x-axis as shown in the following graphs, **CALC minimum** or **maximum** facilities must be used, that will be explained in the following sections.

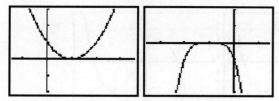

If the left bound and the right bound are not entered correctly, one of the following error messages will appear on the screen.

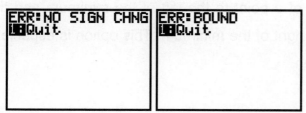

iii. CALC minimum

When **CALC minimum** is selected, the user will be prompted to select the **Left Bound**, and the **Right Bound** of the minimum that is to be found. The left bound must be chosen as a point to the left of the crater of the minimum and the right bound must be chosen as a point to the right of the crater of the minimum. After the **left bound** and the **right bound** are entered correctly, there will be the **guess**ing step which may be passed quickly by pressing the enter key once more.

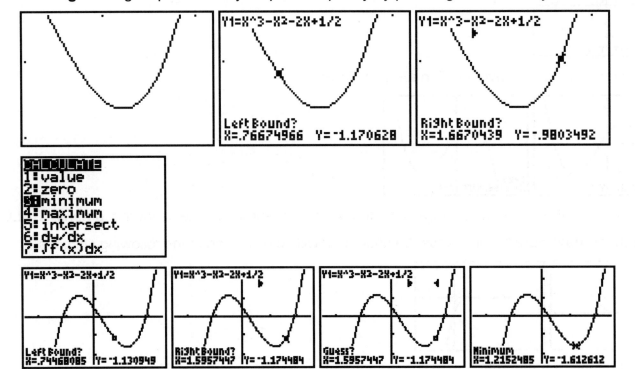

The left bound and the right bound can also be entered manually as shown in the coming graphs. In this case, the left bound will be the x-coordinate of a point to the left of the minimum and the right bound will be the x-coordinate of a point to the right of the minimum. This option is especially useful when the cursor does not appear on the screen.

Advanced Calculation and Graphing Techniques with the TI – 83 Plus Graphing Calculator

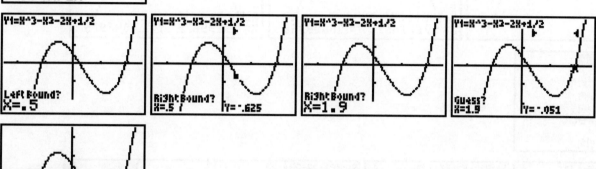

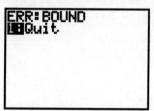

Please note that if the left bound and the right bound are chosen both from the same side of the minimum, there will **NOT** be any error message, and the TI will locate the point lower than the other one, that is the point whose y-coordinate is smaller. The only error message will come up when the left and right bound are selected from reverse sides and it is the following:

iv. CALC maximum

When **CALC maximum** is selected, the user will be prompted to select the **Left Bound**, and the **Right Bound** of the maximum that is to be found. The left bound must be chosen as a point to the left of the crest of the maximum and the right bound must be chosen as a point to the right of the crest of the maximum. After the **left bound** and the **right bound** are entered correctly, there will be the **guess**ing step which may be passed quickly by pressing the enter key once more.

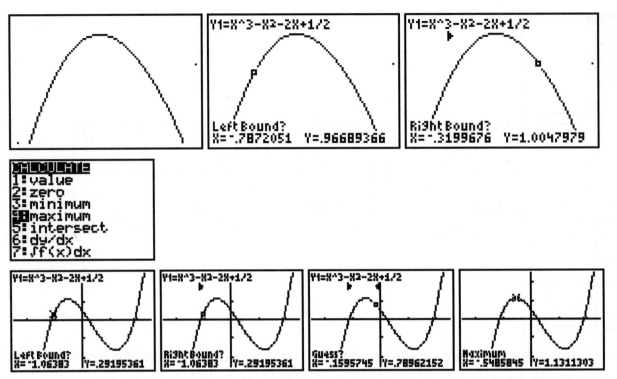

The left bound and the right bound can also be entered manually as shown in the coming graphs. In this case, the left bound will be the x-coordinate of a point to the left of the maximum and the right bound will be the x-coordinate of a point to the right of the maximum. This option is especially useful when the cursor does not appear on the screen.

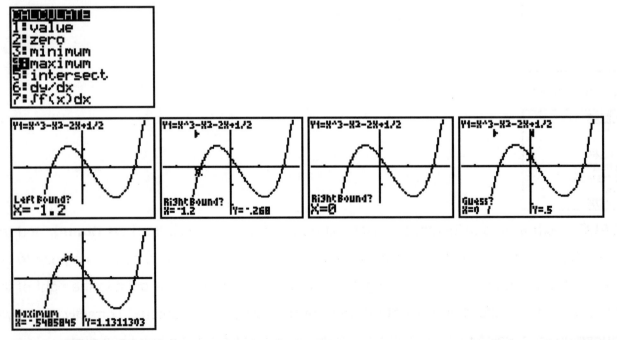

Please note that if the left bound and the right bound are chosen both from the same side of the maximum, there will **NOT** be any error message, and the TI will locate the point upper than the other one, that is the point whose y-coordinate is greater. The only error message will come up when the left and right bound are selected from reverse sides and it is the following:

v. CALC intersect

The **CALC intersect** facility is designed to find the intersection points of functions. In order to be able to use this facility there has to be at least two functions entered in the y-editor.

 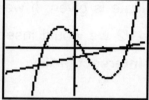

The usage of this facility is simpler than the other ones. The user is prompted to select the first curve and the second curve by pressing the (▼) and the (▲) keys and then guess the intersection point by moving the cursor to the location of the intersection point. When there are only two curves, there is no need to use the (▼) and the (▲) keys to select the curves the curves may be selected rapidly by pressing the enter key twice. The only task left in this case will be to guess the point of intersection by moving the cursor to the point using the (◄) and the (►) keys.

Coming next is a demonstration of how to use this facility to find a desired intersection point. Please note that the only intersection points that can be found are the ones within the viewing window.

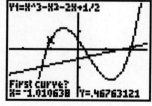 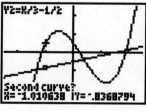 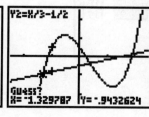 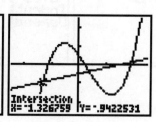

The other points of intersection are as follows:

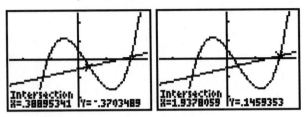

We have seen previously that it is possible to calculate the y-coordinate of a point on the function whose x-coordinate is given by using the **CALC value** facility. The **CALC intersect** facility enables us to find the x-coordinate of a point whose y-coordinate is given. If we would like to find the x coordinates of the points whose y coordinates are −1/2 we would insert y=-1/2 for the second curve and find the intersection points using the **CALC intersect** facility.

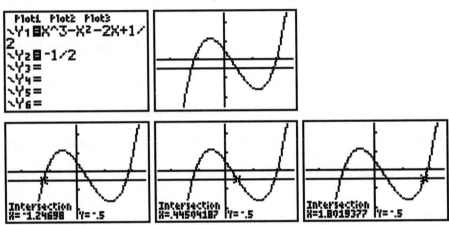

The intersection points above reveal that the x-coordinates of the points whose y-coordinates are −1/2 are −1.24698, .44504187 and 1.8019377.

2.8 Table

When one or more functions are entered in the **Y= Editor**, TI allows the user to construct a table of y values that correspond to the x values whose **starting value** and **increment size** are given by the user. The starting value is the **TblStart** variable and the increment is the **ΔTbl** variable in the **TBLSET**.

(Table Setup) Menu that can be accessed through the key combination of $\boxed{2nd}$ $\boxed{WINDOW}$.

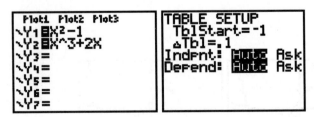

The resulting table can be accessed through the key combination of .

X	Y1	Y2
-1	0	-3
-.9	-.19	-2.529
-.8	-.36	-2.112
-.7	-.51	-1.743
-.6	-.64	-1.416
-.5	-.75	-1.125
-.4	-.84	-.864

X=-1

2.9 Parametric Graphing

Parametric graphing allows the user to plot y versus x when the y- and x- variables are defined in terms of a **parameter "T"** which usually denotes the time variable (for example y- and x- can represent the y- and x-coordinates that define the position of an object at time t). Settings must be changed to the **Par**ametric mode so that the Y= Editor will enable that the relation between y- and x- be both defined in terms of the parameter T. In the **Par**ametric mode the variable T will appear when the (x,T,θ,n) key is pressed. Please note that in the **Func**tion mode the variable X used to appear when the (x,T,θ,n) key was pressed.

For example the following relation

x = 3cos(t)   y = 2sin(t)

must be input as follows (t will be replaced by T):

and the output will be an ellipse:

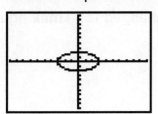

The important issue in the **Par**ametric mode is the fact that the **T** variable in TI is designed to represent the time variable **t** and therefore the default value for **Tmin** in the window settings is 0 (zero) as time is supposed to be always nonnegative.

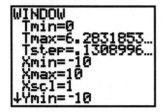

However, **T** does not necessarily have to denote time. For example the parametric equation above could also be given as follows:

x = 3cos(θ)   y = 2sin(θ)

where **θ** is a parameter that does not represent time. On the other hand although the variable used may be t, it may still not represent time, either. In such cases where the free variable does not denote time, leaving **Tmin** as 0 will result in an incorrect and misleading graph that will represent only a portion of the actual graph. Therefore when **T** is not given to represent time, **Tmin** must be changed to **−Tmax**: The default settings for **Tmax** in **Radian** mode is 2π that appears as 6.28… and **Tmin** must be set to -2π that will also appear as -6.28… after enter key is pressed.

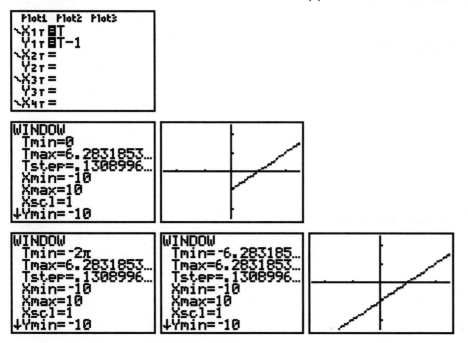

It is also essential to be aware of the fact that the **ZStandard (Zoom Standard)** action will reset **Tmin** back to 0. Therefore after **ZStandard** is performed **Tmin** must be changed to **−Tmax** again when needed.

Advanced Calculation and Graphing Techniques with the TI – 83 Plus Graphing Calculator

2.10 Polar Graphing

Polar graphing allows the user to plot **r** versus **θ** where **r** denotes **radius**, **θ** denotes **angle** and **(r,θ)** represents the **polar coordinates**. Settings must be changed to the **Pol**ar mode so that the Y= Editor will enable that **r** be defined in terms of the parameter **θ**. In the **Pol**ar mode the variable **θ** will appear when the (x,T,θ,n) key is pressed.

For example the input and output for the following relation

$r = 1 - 2\cos(\theta)$

will be as follows:

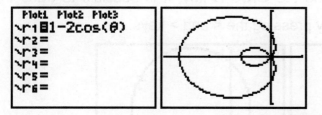

2.11 Graphing of Conics

In the TI-83 Plus there is the **APPS** key that enables access to several useful applications. These applications are either built in or downloadable from the TI site **education.ti.com**. A few of these applications are especially designed to make the lives of a college bound high school student easier. One of these applications is named as **Conics**. TI enables the graphing of conic sections namely the circle, ellipse, parabola, and hyperbola accessible through the combination of **APPS Conics**.

Circle

The first sub-menu in the **APPS Conics** menu is the Circle.

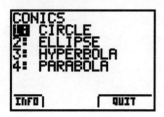

This sub menu makes it possible to graph a circle given in either of the following two forms.

i. $(x-H)^2+(y-K)^2=R^2$

ii. $Ax^2+Ay^2+BX+CY+D=0$

$(x-H)^2+(y-K)^2=R^2$: In this form (H,K) is the center and R is the radius of the given circle. In order to graph the circle whose equation is given in this form, the circle equation must initially be converted to this form by completing squares and the H, K, and R values obtained must be entered prior to pressing the **GRAPH** key. Please note that R cannot be a negative entry.

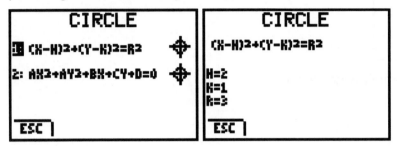

After entering the H, K, and R values and pressing the **GRAPH** key, it is possible to trace the points of the circle in a counter clockwise fashion by pressing the < and > keys.

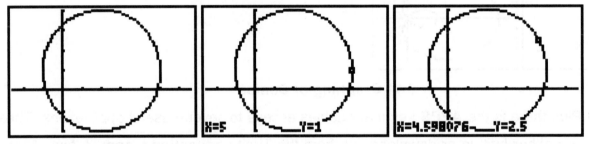

$Ax^2+Ay^2+BX+CY+D=0$: In order to graph the circle whose equation is given in the above form, A, B, C, and D must be entered prior to pressing the **GRAPH** key. Please note that A must be nonzero and that the value B^2+C^2-4AD must be positive for a real circle.

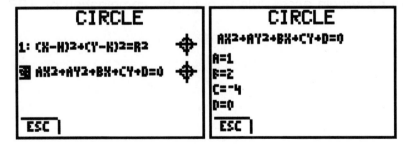

Advanced Calculation and Graphing Techniques with the TI – 83 Plus Graphing Calculator

Unauthorized copying or reuse of any part of this page is illegal.

After entering the A, B, C, D values and pressing the **GRAPH** key, it is possible to trace the points of the circle in a counter clockwise fashion by pressing the < and > keys.

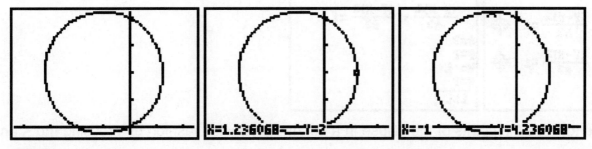

Ellipse

The second sub-menu in the **APPS Conics** menu is the Ellipse.

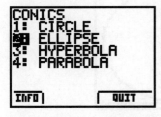

Depending on whether the ellipse is an x-ellipse or a y-ellipse, the ellipse must be converted to either of the following forms by using the method of completing squares:

i. x-ellipse: $\dfrac{(x-H)^2}{A^2} + \dfrac{(y-K)^2}{B^2} = 1$

ii. y-ellipse: $\dfrac{(x-H)^2}{B^2} + \dfrac{(y-K)^2}{A^2} = 1$

(H,K) is the center of the ellipse; A is the semi-major axis length and B is the semi-minor axis length of the ellipse in both of the above forms. Please note that A and B are positive and A is greater than B. Entries violating the assumption that A>B>0 will result in an error message illustrated as follows:

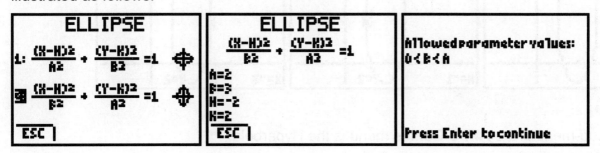

An x-ellipse is given in the following example:

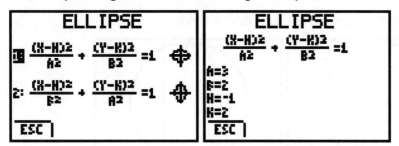

It is possible to trace the points of the ellipse in a counter clockwise fashion by pressing the < and > keys.

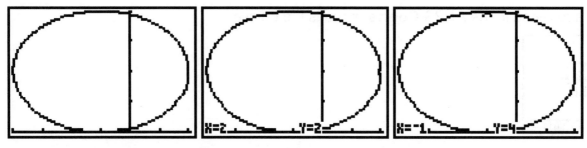

A y-ellipse is given in the following example:

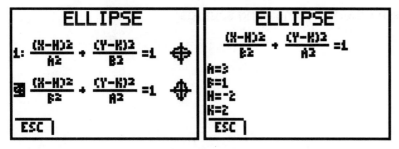

It is possible to trace the points of the ellipse in a counter clockwise fashion by pressing the < and > keys.

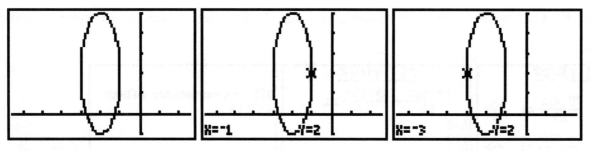

Hyperbola

The third sub-menu in the **APPS Conics** menu is the Hyperbola.

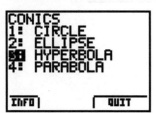

Advanced Calculation and Graphing Techniques with the TI – 83 Plus Graphing Calculator

Depending on whether the hyperbola is an x-hyperbola or a y-hyperbola, the hyperbola must be converted to either of the following forms by using the method of completing squares:

i. x: hyperbola: $\dfrac{(x-H)^2}{A^2} - \dfrac{(y-K)^2}{B^2} = 1$

ii. y hyperbola: $\dfrac{(y-K)^2}{A^2} - \dfrac{(x-H)^2}{B^2} = 1$

(H,K) is the center of the hyperbola; A is the semi-transversal axis length and B is the semi-conjugate axis length of the hyperbola in both of the above cases.

An x-hyperbola is given in the following example:

It is possible to trace the points of the ellipse in a counter clockwise fashion by pressing the < and > keys.

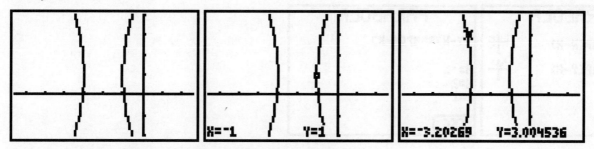

A y-hyperbola is given in the following example:

Advanced Calculation and Graphing Techniques with the TI – 83 Plus Graphing Calculator

It is possible to trace the points of the ellipse in a counter clockwise fashion by pressing the < and > keys.

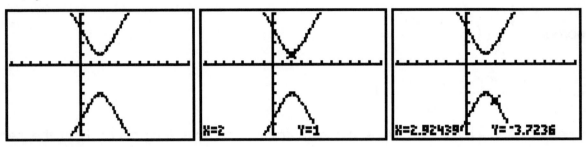

Parabola

The fourth sub-menu in the **APPS Conics** menu is the Parabola.

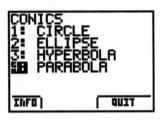

If the arms of the parabola open rightward or leftward then it is in the following form:

$(y-K)^2 = 4P(x-H)$

An example is given as follows:

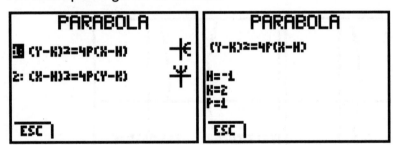

It is possible to trace the points of the parabola by pressing the < and > keys.

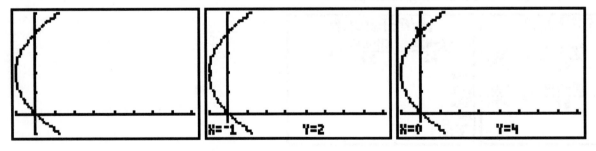

If the arms of the parabola open upward or downward then it is in the following form:

$(x-H)^2 = 4P(y-K)$

An example is given as follows:

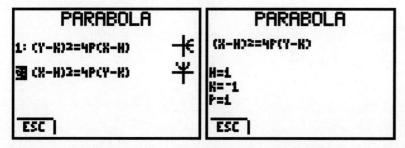

It is possible to trace the points of the parabola by pressing the < and > keys.

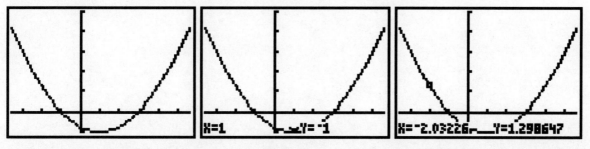

In either form (H,K) is the vertex and |P| is the center to focus distance of the parabola.

NOTES

CHAPTER 3.
THE METHOD

Advanced Calculation and Graphing Techniques with the TI – 83 Plus Graphing Calculator

Solving Polynomial or Algebraic Equations

When solving a polynomial or algebraic equation in the form **f(x)=g(x),** perform the following steps:

i.   Write the equation in the form: **f(x)-g(x)=0**.
ii. Plot the graph of **y=f(x)-g(x)**.
iii. Find the x-intercepts using the **Calc Zero** of TI-83 Plus. However when the graph seems to be tangent to the x-axis at a certain point, you may use the **Calc Min** or **Calc Max** facilities but you should make sure that the y-coordinate of the minimum or maximum point is zero.
iv. Any value like **-6.61E -10** or **7.2E -11** can be interpreted as 0 as they mean **-6.6x10^{-10}** and **7.2x10^{-11}** respectively.

Solving Absolute Value Equations

When solving an absolute value equation in the form **f(x)=g(x),** perform the following steps:

i.   Write the equation in the form: **f(x)-g(x)=0**. Whenever absolute values have to be involved, replace **| f(x) |** by **abs(f(x))**
ii. Plot the graph of **y=f(x)-g(x)**.
iii. Find the x-intercepts using the **Calc Zero** of TI-83 Plus. However when the graph seems to be tangent to the x-axis at a certain point, you may use the **Calc Min** or **Calc Max** facilities but you should make sure that the y-coordinate of the minimum or maximum point is zero.
iv. Any value like **-6.61E -10** or **7.2E -11** can be interpreted as 0 as they mean **-6.6x10^{-10}** and **7.2x10^{-11}** respectively.

Solving Exponential and Logarithmic Equations

When solving an exponential or logarithmic equation in the form **f(x)=g(x)**, perform the following steps:

i.   Write the equation in the form: **f(x)-g(x)=0**.
ii. Whenever logarithms have to be involved, replace **log $_{f(x)}$ g(x)** by $\frac{\log g(x)}{\log f(x)}$ or by $\frac{\ln g(x)}{\ln f(x)}$. Whenever exponentials have to be involved, replace **f(x) $^{g(x)}$** by **f(x)^g(x)**; $\sqrt[n]{g(x)}$ by **g(x)^(1/n)**; and $\sqrt[n]{g(x)^m}$ by **g(x)^(m/n)**; **exp(f(x))** must be interpreted as **e $^{f(x)}$**.

iii. Plot the graph of **y=f(x)-g(x)**.

iv. Find the x-intercepts using the **Calc Zero** of TI-83 Plus. However when the graph seems to be tangent to the x-axis at a certain point, you may use the **Calc Min** or **Calc Max** facilities but you should make sure that the y-coordinate of the minimum or maximum point is zero.

v. Any value like **-6.61E -10** or **7.2E -11** can be interpreted as 0 as they mean **-6.6x10 $^{-10}$** and **7.2x10 $^{-11}$** respectively.

Solving System of Linear Equations

In order to solve a system of linear equations, the coefficients of the linear system are entered in two separate matrices A and B followed by the simple algebraic operation of $A^{-1} * B$. The matrix menu is accessed by pressing the [x^{-1}] key. The matrix entries are entered through the **EDIT** sub menu where the dimensions of the matrix followed by the matrix entries must be input. The matrices that are entered will be used through the **NAMES** sub menu later on. In the following example A is a 2 by 2 matrix and B is a 2 by 1 matrix.

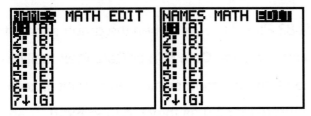

Solving Trigonometric Equations

When solving a trigonometric equation in the form **f(x)=g(x)**, perform the following steps:

i. Write the equation in the form: **f(x)-g(x)=0**.

ii. While writing the trigonometric expressions please observe the rules given in parts 1.17 and 1.18 that involve the trigonometric functions that are built in TI or otherwise.

iii. Plot the graph of **y=f(x)-g(x)**.

iv. Set the **angle mode to radians or degrees** depending on which angle measure is used in the question. If no degree signs (like in 90°) are used then the mode should be radians. However when exact values are required you may wish to solve the equation in degrees and convert the answer to radians using the following formula: $\dfrac{R}{\pi} = \dfrac{D}{180°}$. In such a case

Advanced Calculation and Graphing Techniques with the TI – 83 Plus Graphing Calculator

finding the answer in radians and then trying to find which answer choice matches this answer can also be an option; while doing so you may directly replace π with 180°

v. If x is limited to a certain interval then set **Xmin**; **Xmax** and **Xscl** accordingly. For example, if x is an acute angle and the angle mode is degrees, then **Xmin** must be set to 0°; **Xmax** must be set to 90° and **Xscl** must be set so that the grigding of the x-axis will be made properly In such a case **Xscl** being 30° would be fine. If x is an acute angle and the angle mode is radians, then **Xmin** must be set to 0; **Xmax** must be set to $\pi/2$ and **Xscl** may be set to 1.

vi. When only sines and cosines are involved, **ZoomFit** option may give a clearer graph. However, since only the x-intercepts are required, the window setting parameters **Ymin= -1** and **Ymax = 1** can give a clear view of the zeros.

vii. Find the x-intercepts using the **Calc Zero** of TI-83 Plus. However when the graph seems to be tangent to the x-axis at a certain point, you may use the **Calc Min** or **Calc Max** facilities but you should make sure that the y-coordinate of the minimum or maximum point is zero.

viii. Any value like **-6.61E -10** or **7.2E -11** can be interpreted as 0 as they mean **-6.6x10^{-10}** and **7.2x10^{-11}** respectively.

Solving Inverse Trigonometric Equations

When solving an inverse trigonometric equation in the form **f(x)=g(x)**, perform the following steps:

i. Write the equation in the form: **f(x)-g(x)=0**.

ii. While writing the trigonometric expressions please observe the rules given in parts 1.17 and 1.18 that involve the trigonometric functions that are built in TI or otherwise.

iii. Plot the graph of **y=f(x)-g(x)**.

iv. Set the **angle mode to radians or degrees** depending on which angle measure is used in the question. If no degree signs (like in 90°) are used then the mode should be radians. However when exact values are required you may wish to solve the equation in degrees and convert the answer to radians using the following formula: $\dfrac{R}{\pi} = \dfrac{D}{180°}$. In such a case finding the answer in radians and then trying to find which answer choice matches this answer can also be an option; while doing so you may directly replace π with 180°

Advanced Calculation and Graphing Techniques with the TI – 83 Plus Graphing Calculator

v. Find the x-intercepts using the **Calc Zero** of TI-83 Plus. However when the graph seems to be tangent to the x-axis at a certain point, you may use the **Calc Min** or **Calc Max** facilities but you should make sure that the y-coordinate of the minimum or maximum point is zero.

vi. Any value like **-6.61E -10** or **7.2E -11** can be interpreted as 0 as they mean **-6.6x10^{-10}** and **7.2x10^{-11}** respectively.

Solving Inequalities

When solving an inequality in the form **f(x)<g(x)**, or **f(x)≤g(x)**, or **f(x)>g(x)**, or **f(x)≥g(x)** perform the following steps:

i. Write the inequality in the form: **f(x)-g(x)<0 or f(x)-g(x)≤0 or f(x)-g(x) >0 or f(x)-g(x)≥0.**

ii. Plot the graph of **y=f(x)-g(x)**.

iii. Find the x-intercepts using the **Calc Zero** of TI-83 Plus. However when the graph seems to be tangent to the x-axis at a certain point, you may use the **Calc Min** or **Calc Max** facilities but you should make sure that the y-coordinate of the minimum or maximum point is zero.

iv. Any value like **-6.61E -10** or **7.2E -11** can be interpreted as 0 as they mean **-6.6x10^{-10}** and **7.2x10^{-11}** respectively.

v. The solution of the inequality will be the set of values of x for which the graph of f(x)-g(x) lies below the x axis if the inequality is in one of the forms **f(x)-g(x)<0** or **f(x)-g(x)≤0**. The solution of the inequality will be the set of values of x for which the graph of f(x)-g(x) lies above the x axis if the inequality is in one of the forms **f(x)-g(x)>0** or **f(x)-g(x)≥0**. If ≤ or ≥ symbols are involved, then the x-intercepts are also in the solution set.

vi. Please note that the x-values that correspond to asymptotes are never included in the solution set.

Solving Trigonometric Inequalities

When solving a trigonometric inequality in the form **f(x)<g(x)**, or **f(x)≤g(x)**, or **f(x)>g(x)**, or **f(x)≥g(x)** perform the following steps:

i. Write the inequality in the form: **f(x)-g(x)<0 or f(x)-g(x)≤0 or f(x)-g(x) >0 or f(x)-g(x)≥0.**

ii. While writing the trigonometric expressions please observe the rules given in parts 1.17 and 1.18 that involve the trigonometric functions that are built in TI or otherwise.

iii. Plot the graph of **y=f(x)-g(x)**.

Advanced Calculation and Graphing Techniques with the TI – 83 Plus Graphing Calculator

Unauthorized copying or reuse of any part of this page is illegal.

iv. Set the **angle mode to radians or degrees** depending on which angle measure is used in the question. If no degree signs (like in 90°) are used then the mode should be radians. However when exact values are required you may wish to solve the equation in degrees and convert the answer to radians using the following formula $\frac{R}{\pi} = \frac{D}{180°}$. In such a case finding the answer in radians and then trying to find which answer choice matches this answer can also be an option; while doing so you may directly replace π with 180°

v. If x is limited to a certain interval then set **Xmin**; **Xmax** and **Xscl** accordingly. For example, if x is an acute angle and the angle mode is degrees, then **Xmin** must be set to 0°; **Xmax** must be set to 90° and **Xscl** must be set so that the grigding of the x-axis will be made properly In such a case **Xscl** being 30° would be fine. If x is an acute angle and the angle mode is radians, then **Xmin** must be set to 0; **Xmax** must be set to $\pi/2$ and **Xscl** may be set to 1.

vi. When only sines and cosines are involved, **ZoomFit** option may give a clearer graph. However, since only the x-intercepts are required, the window setting parameters **Ymin= -1** and **Ymax = 1** can give a clear view of the zeros.

vii. Find the x-intercepts using the **Calc Zero** of TI-83 Plus. However when the graph seems to be tangent to the x-axis at a certain point, you may use the **Calc Min** or **Calc Max** facilities but you should make sure that the y-coordinate of the minimum or maximum point is zero.

viii. Any value like **-6.61E -10** or **7.2E -11** can be interpreted as 0 as they mean **-6.6x10^{-10}** and **7.2x10^{-11}** respectively.

ix. The solution of the inequality will be the set of values of x for which the graph of f(x)-g(x) lies below the x axis if the inequality is in one of the forms **f(x)-g(x)<0** or **f(x)-g(x)≤0**. The solution of the inequality will be the set of values of x for which the graph of f(x)-g(x) lies above the x axis if the inequality is in one of the forms **f(x)-g(x)>0** or **f(x)-g(x)≥0**. If ≤ or ≥ symbols are involved, then the x-intercepts are also in the solution set.

x. Please note that the x-values that correspond to asymptotes are never included in the solution set.

Advanced Calculation and Graphing Techniques with the TI – 83 Plus Graphing Calculator

Finding Maxima and Minima

When solving for the maximum and/or minimum points of a function **f(x)** perform the following steps:

i. If x is limited to a certain interval then set **Xmin**; **Xmax** and **Xscl** accordingly, otherwise use Zstandard facility while graphing **y=f(x)**.

ii. Use the **Calc Min** or **Calc Max** facilities to find the minimum and maximum point(s). However if the minimum or maximum points are at one or both of the ends of the interval, then find these points by using the **Calc Value** facility; while doing so, use the x-coordinates of the endpoints of the interval.

iii. Any value like **-6.61E -10** or **7.2E -11** can be interpreted as 0 as they mean -6.6×10^{-10} and 7.2×10^{-11} respectively.

Finding Domains and Ranges

When finding the domain and range of a function **f(x)**, graph the function and simply find the set of x values for which f(x) is plotted. You may perform the following steps:

i. If x is limited to a certain interval then set **Xmin**; **Xmax** and **Xscl** accordingly, otherwise use Zstandard facility while graphing **y=f(x)**.

ii. Use the **Calc Zero, Calc Value, Calc Min**, or **Calc Max** facilities to find the zeros, minima and maxima.

iii. When asymptotes or discontinuities are involved, you may use the **TBLSET** and **TABLE** facilities to find the set of x values for which f(x) is undefined or not continuous.

iv. Any value like **-6.61E -10** or **7.2E -11** can be interpreted as 0 as they mean -6.6×10^{-10} and 7.2×10^{-11} respectively.

Exploring Evenness and Oddness

When finding whether a function **f(x)** is **even, odd** or **neither**, graph the function and simply check the symmetry.

i. If f(x) is symmetric in the y-axis, then it is even.

ii. If f(x) is symmetric in the origin, then it is odd.

iii. If f(x) is not symmetric in the y-axis or the origin then it is neither even nor odd.

Advanced Calculation and Graphing Techniques with the TI – 83 Plus Graphing Calculator

Graphs of Trigonometric Functions

(i) Most of the time one or more of the following are required concerning the graphs of the trigonometric functions. In order to find them all it is usually enough to find two adjacent maxima and the minimum point in between.

Period = The x-distance between two identical points in a periodic function; for example two adjacent maxima, minima or zeros.

Frequency = 1 / Period

Amplitude = (Ymax – Ymin) / 2

Offset = (Ymax + Ymin) / 2

Axis of wave equation: y = Offset

(ii) **y-intercept** is the point whose x-coordinate is zero.

(iii) Use the window, **Calc Min**, **Calc Max**, **Calc Value**, and **Calc Zero** facilities in order to perform the above calculations.

The Greatest Integer Function

The greatest integer function f(x) = [x] = [|x|] means "The greatest integer less than or equal to x". Mathematical definition for the greatest integer function is as follows:

f(x) = k if k ≤ x<k+1 and k=integer ⇒ f(x) = [x]

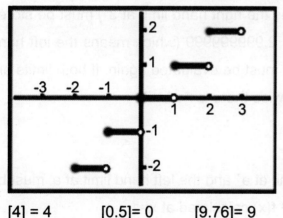

[4] = 4   [0.5]= 0   [9.76]= 9   [-3]= -3   [-8.67]= -9   [-0.32]= -1

TI Usage

y=int(x) and style must be set to dot

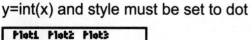

Advanced Calculation and Graphing Techniques with the TI – 83 Plus Graphing Calculator

Parametric Graphing

Please review section 2.9.

Polar Graphing

Please review section 2.10.

Limits

(i) For a function to have a limit for a given value of x=a, the right hand limit at a^+ and the left hand limit at a^- must be the same and each limit must be equal to a real number L other than infinity.

Existence of Limit: If $\lim_{x \to a^+} f(x) = \lim_{x \to a^-} f(x) = L$ and $L \in R$ then $\lim_{x \to a} f(x) = L$

(ii) Limit for a certain value of x or limit at infinity can be calculated by using the **STO**re facility of TI. What must be done is simply to store a value in x and calculate the value of the expression for this x-value.

(iii) ∞ can be replaced by 100,000,000,000; and -∞ can be replaced by -100,000,000,000.

(iv) Limit at a value of x other than ±∞ must be calculated as follows: If for example the limit at x=3 will be calculated, 3.000000001 (which means the right hand limit at 3^+) must be stored in x and the expression must be evaluated; then 2.999999999 (which means the left hand limit at 3^-) must be stored in x and the expression must be evaluated again. If both limits are the same, say L, then the limit is equal to L, otherwise there is no limit.

Continuity

For a function to be continuous at x=a, the right hand limit at a^+ and the left hand limit at a^- must be the same and the limit must also be equal to the value of f(x) calculated at x=a.

If $\lim_{x \to a^+} f(x) = \lim_{x \to a^-} f(x) = f(a)$ then f(x) is continuous at x=a.

Horizontal and Vertical Asymptotes

$f(x) = \dfrac{P(x)}{Q(x)}$ where P(x) and Q(x) are both polynomial functions.

Advanced Calculation and Graphing Techniques with the TI – 83 Plus Graphing Calculator

(i) **Zero:** If $P(x_o)=0$ and $Q(x_o) \neq 0$ then $f(x)$ has a zero at $x=x_o$.

(ii) **Hole:** If $P(x_o)=0$ and $Q(x_o)=0$, and the multiplicity of x_o is the same in both polynomials, then $f(x)$ has a hole at $x=x_o$

(iii) **Vertical asymptote:** If $P(x_o) \neq 0$ but $Q(x_o)=0$, then $f(x)$ has a vertical asymptote at $x=x_o$

(iv) **Horizontal asymptote:** If the limit of $\frac{P(x)}{Q(x)}$ equals b as x goes to $\pm \infty$ then y=b is the horizontal asymptote.

Complex Numbers

(i) Please review section 1.16.

(ii) Calculator mode must be set to **a+bi** where i is the imaginary number that has the following properties: $i^2 = -1$ and $i = \sqrt{-1}$. The required functions can be found at the **MATH CPX** menu:

(iii) **abs**(a+bi)=$\sqrt{a^2+b^2}$; **real**(a+bi)=a; **imag**(a+bi)=b; **conj**(a+bi)=a–bi.

(iv) cis(x)=cos(x)+i.sin(x)

(v) e^{ix}= cos(x)+i.sin(x)

Permutations and Combinations

Please review section 1.21. and note the following:

n! = n.(n-1).(n-2).(n-3)...3.2.1

P(n,r): Number of permutations of r elements chosen from n elements; P(n,r)=$\frac{n!}{(n-r)!}$ and

C(n,r): Number of combinations of r elements chosen from n elements; C(n,r)= $\frac{n!}{(n-r)!.r!}$

where $r \leq n$. The required functions can be found at the **MATH PRB** menu.

NOTES

CHAPTER 4.

SAMPLE TI PROBLEMS

(ANSWERS ON PAGE 180; SOLUTIONS ON PAGE 69)
4.1 Polynomial Equations

1. $P(x)= 3x^3-5x^2+6x-3$

   The zero of the above polynomial lies between two consecutive integers. What are these integers?

2. $y= 3x^2-4x-5$

   What is the positive zero of the above function correct to the nearest hundredth?

3. Is $x-99$ a factor of the following polynomial?
   $P(x)= 2x^4-200x^3+194x^2+400x-394$

4. Find all real zeros of the following polynomial:
   $P(x)= 2x^6-2x^5-8x^4-2x^3+10x^2+16x+8$

5. What is the least positive integer greater than the zero of the following polynomial?
   $P(x)= -\dfrac{3}{2}x^3 - x^2 - 2x + 3$

6. What are the real roots of the following function?
   $f(x)= -2x^4-4x^3+6x^2-4x+7$

7. $P(x)= 3x^4 - x^3 + 2x^2 + 5x - 1$

   How many positive and negative real zeros does the above polynomial have?

8. $x^2 + x + 2 = 0$

   What is the nature of the roots of the above equation?

9. $P(x)= 2x^3 + x^2 + 3x - 5$

   How many positive and negative real zeros does the above polynomial have?

10. Find the positive rational root of the following equation: $2x^3 - 5x^2 + 14x = 35$

11. What is the absolute difference between the zeros of the following function?
    $f(x)= 7x^2 + 11.5x - 25$

12. What is the sum of the zeros of the following parabola?
    $y= 3x^2 - 7x - 5$

13. What are the zeros of $y= 3x^2 + x - 4$?

14. $f(x)= 6x^2 + 12x - 3$; $f(q)=0 \Rightarrow$ What is one value of q?

15. Find the sum of the roots of $6x^3+8x^2 - 8x=0$

16. $P(x)= 2x^2+3x+1$; $P(a)= 7 \Rightarrow a=?$

17. What is the product of the roots of the following equation $(x-\sqrt{3})(x^2 - ex - \pi) = 0$?

18. $f(x)= 5x^2-7$. Find sum of the zeros of $f(x)$.

Advanced Calculation and Graphing Techniques with the TI – 83 Plus Graphing Calculator

19. $P(x) = x^3 + 6x - 14$ has a zero between which two consecutive integers?

20. $f(x) = x^2 - 9$

    $(f \circ f)(x) = 0 \Rightarrow$ What are the real values of x?

21. $2x^4 + 3x^3 + 2x - 1 = 0$

    Find nature of the roots.

22. Is $3x+1$ a factor of $2x^3 + 4x^2 - 4x - 3$?

23. Find the number of the positive real zeros of the following equation:

    $x^4 + 2x^3 - 4x^2 - 5x = 0$

24. Find product of the real roots of the following equation:

    $x^4 - 3x^3 - 72x^2 - 3x - 18 = 0$

4.2 Algebraic Equations

1. $f(x) = \sqrt{3x+4}$

    $g(x) = x^3$

    If is given what $(f \circ g)(x) = (g \circ f)(x)$, then what is x?

2. $f(x) = \sqrt{-x^3 + 4x}$

    $g(x) = 4x$

    What is the sum of the roots of the equation $f(x) = g(x)$?

3. $a \# b = a^b - b^a$

    If $3 \# k = k \# 2$ then $k = ?$

4. $5x^{4/3} = 2 \Rightarrow x = ?$

5. Find sum of the roots of: $2x - \dfrac{5}{x} + 2 = 0$

4.3 Absolute Value Equations

1. $|3x-1| = 4x+6$

    How many numbers are there in the solution set of the above equation?

2. $|x-3| + |2x+1| = 6 \Rightarrow x = ?$

3. $|3x-5| = 4 \Rightarrow x = ?$

4. $|4x+6| = 3x+4 \Rightarrow x = ?$

5. $\dfrac{|x-3|}{x} = 4 \Rightarrow x = ?$

Advanced Calculation and Graphing Techniques with the TI – 83 Plus Graphing Calculator

4.4 Exponential and Logarithmic Equations

1. $\log_4 x \cdot \log_5 6 = 7 \Rightarrow x = ?$

2. $A = e^{Bt}$

    $A = 1000, T = 4, B = ?$

3. $f(x) = \exp(x)$

    $(\exp(x) = e^x)$

    If $h(x) = f(-x) + f^{-1}(-x)$ then $h(-2) = ?$

4. If $\log x = \dfrac{3}{4}$ then $\log(1000x^2) = ?$

5. $\left.\begin{array}{l}\log_3 x = \sqrt{5} \\ \log_5 y = \sqrt{3}\end{array}\right\} x \cdot y = ?$

6. $\log_3 2 = x \cdot \log_6 5 \Rightarrow x = ?$

7. $2^{x+3} = 3^x \Rightarrow x = ?$

8. $\log_x 3 = \log_4 x \Rightarrow$ What is the sum of the roots of this equation?

9. $f(x) = 3.5^x + 1;\ f^{-1}(10) = ?$

10. $3.281^x = 4.789^y \Rightarrow \dfrac{x}{y} = ?$

4.5 System of Linear Equations, Matrices and Determinants

1. $\left.\begin{array}{l}x + 3y = 7 \\ 12x - 2y = 8\end{array}\right\} \dfrac{x}{y} = ?$

2. $\left.\begin{array}{l}x + y + z = 6 \\ 2x - y + 3z = 9 \\ 3x + y - 4z = -7\end{array}\right\} x^2 + y^2 + z^2 = ?$

3. If $A = \begin{bmatrix} 2 & 3 \\ -1 & 5 \end{bmatrix}$, $B = \begin{bmatrix} 0 & -2 \\ -3 & 5 \end{bmatrix}$,

    Find

    i. $A + B$

    ii. $3A - 2B$

    iii. AB

    iv. BA

4. Find the determinant and inverse of the matrix $\begin{bmatrix} 1 & 3 & -1 \\ -2 & 4 & 1 \\ 0 & 0 & 2 \end{bmatrix}$

4.6 Trigonometric Equations

1. How many solutions does the following equation have between 0° and 360°?

    $\sec^2 x - \dfrac{\sin x}{\cos x} = 1$

2. $\cos(33°) = \tan x° \Rightarrow x=?$ (x is an acute angle)

3. $0 < x < \dfrac{\pi}{4}$ and $\tan(4x)=3$. What is x and what is tan x?

4. $\sin(120°-n) = \sin 50°$ and n is an acute angle $\Rightarrow n=?$

5. What is the sum of the two least positive solutions of the following equation?

    $\sin(10x) = -\cos(10x)$

6. $\cos(2x) = 2\sin(90°-x)$. What are all possible values of x between 0° & 360°?

7. $\dfrac{1}{4}\sin^2(2x) + \sin^2(x) + \cos^4(x) = 1$

    If x is positive and less than 2π, how many different values can x have?

8. $\dfrac{1}{\cot(5x)} = -2$

    What is the smallest positive value for x?

9. $\dfrac{8\sin(2\theta)}{1-\cos(2\theta)} = \dfrac{4}{3}$ and θ is between 0° and 180°. What is θ?

10. $\dfrac{\sin x + \cos 36°}{\cos \dfrac{4\pi}{3} - \sin(-90°)} = 0$ and x is between 90° and 270° $\Rightarrow x=?$

11. $\dfrac{\sin\theta}{\cos\theta - 1} = -\sqrt{3} \Rightarrow$ If θ is an acute angle, $\theta=?$

12. $2\sin x + \cos(2x) = 2\sin^2 x - 1$ and $0 \le x < 2\pi \Rightarrow x=?$

13. $\cos(130°-2x) = \sin(70°-3x)$ and x is an acute angle. What is x?

14. x is in quadrant 3 and $\cot(120°-x) = \dfrac{1}{\tan x} \Rightarrow x=?$

15. $\left.\begin{array}{l} \dfrac{\sin(2\theta)}{2} = \dfrac{1}{4} \\ 0° \le \theta < 360° \end{array}\right\}$ What is θ?

16. $\left.\begin{array}{l} \sec\theta \cdot \csc\theta = 4 \\ 0° \le \theta < 360° \end{array}\right\}$ $\theta=?$

17. $\left.\begin{array}{l} 0° \le x < 90° \\ \tan(4x) = 1 \end{array}\right\}$ $x=?$

18. $\tan(6x) = \sqrt{3}$ and x is an acute angle $\Rightarrow$ x=?

19. $\left.\begin{array}{l}\dfrac{\sqrt{3}}{2}\cos x + \dfrac{1}{2}\sin x = 1 \\ 0 \le x < 2\pi \end{array}\right\} \Rightarrow$ x=?

20. $2\sin^2 x = 3(1+\cos x) - \dfrac{1}{2}$ and x is in 3rd quadrant. What is x in radians?

21. $\cos x \cos 45° - \sin x \sin 45° = -1$ and x is an obtuse angle $\Rightarrow$ x=?

22. $\left.\begin{array}{l}\sin x \sec x = \sqrt{3} \\ 0 \le x < 2\pi \end{array}\right\} \Rightarrow$ x=?

4.7 Inverse Trigonometric Equations

1. Solve for x: $\cos^{-1}(2x - 2x^2) = \dfrac{2\pi}{3}$

2. $\sin^{-1}(x) = 3\text{ Arccos} x \Rightarrow$ x=?

3. $\left.\begin{array}{l} A = \text{Arctan}\left(\dfrac{-5}{12}\right) \\ A + B = 300° \end{array}\right\} \Rightarrow$ B=?

4.8 Polynomial, Algebraic and Absolute Value Inequalities

Find the solution sets of the following:

1. $x^2 - 8x + 7 < 0$

2. $\dfrac{x}{x-3} > 4$

3. $\dfrac{|x-2|}{x} > 3$

4. $f(x) = x + \sqrt{2x+1}$ and $f(x) \le 4$.

5. $x(x-1)(x+2)(x-3) < 0$

6. $x(x-2)(x+1) > 0$

7. $x^2(x-2)(x+1) \ge 0$

8. $\dfrac{x+2}{x} < 4$

9. $4x^2 - x < 3$

10. $\dfrac{(x+1)^2}{x^2} > 0$

11. $|2x + 5| \ge 3$

12. $|x-2| \leq 1$

13. $x^2+12 < 7x$

14. In which quadrants are the points that satisfy the following system of inequalities?

    $y < -(x-2)^2 - 1$

    $y \geq 2x - 7$

4.9 Trigonometric Inequalities

1. Find the solution set of the following inequality: $x < \cos x$

2. $\sin(2x) > \sin x$

   Find the set of values of x that satisfy the above inequality in the interval $0 < x < 2\pi$.

3. If x is between 0 and 2π, what will be the set of x values for which $\sin x < \cos x$?

4. $\cos(2x) \geq \cos x$

   Find the set of values of x that satisfy the above inequality in the interval $0 \leq x \leq 360°$.

4.10 Maxima and Minima

1. $f(x) = 2x^2 + 1$ is defined in the interval $-3 \leq x \leq 3$. find minimum value of f(x).

2. $f(x) = |3x+1| - 1$ Find minimum value of f(x).

3. $f(x) = -|x| + 3$ and $-2 \leq x \leq 4$ Find minimum value of f(x) and the x value where this minimum occurs.

4. $y = \sqrt[3]{9-x^2}$. Find maximum value of y.

4.11 Domains and Ranges

1. Find the domain of $f(x) = \log \sqrt{2x^2 - 15}$.

2. Find domain and range of the function $y = x^{-4/3}$

3. Find the domain and range of the function $f(x) = 4 - \sqrt{2x^3 - 16}$

4. $f(x) = 2x^2 + 5x + 2$ and $g(x) = 4x^2 - 4$.

   In order that $\left(\dfrac{f}{g}\right)(x)$ be a function what must be excluded from the domain?

5. Find range of $y = 8 - 2x - x^2$

6. $f(x) = \log(\sin x)$. Find domain of f(x).

7. $f(x) = \dfrac{3x+4}{x+2}$. Find domain and range of f(x)

8. $f(x) = \dfrac{x+1}{2x-2}$. What value(s) must be excluded from the domain of f(x) and what is the range of f(x)?

9. What is the domain and range of $y = \dfrac{x^2 - 4}{x^2 - 2x}$?

10. Domain of f(x) is given by $x^2 + 3x - 4 < 0$ and $f(x) = x^2 + 4x + 5$. Find range of f(x).

11. Find domain and range of $y = \sqrt{x^2 - 9}$.

12. Find domain and range of $y = \sqrt{9 - x^2}$.

4.12 Evenness And Oddness

State whether each of the following functions are even, odd, or neither.

1. $f(x) = \dfrac{1}{\sec x}$
2. $f(x) = \cos x$
3. $f(x) = \dfrac{1}{\csc(x)}$
4. $f(x) = \sin x$
5. $f(x) = \sin x + 1$
6. $f(x) = \dfrac{1}{x}$
7. $f(x) = |x|$
8. $f(x) = \log(x^2)$
9. $f(x) = -x^2 + \sin x$
10. $f(x) = x^4 - 3x^2 + 5$
11. $f(x) = 3x^3 + 5$
12. $f(x) = 12x^6 + 4x^4 - 13x^2$
13. $f(x) = -x^5 - 8x^3 + 12x$
14. $f(x) = x^3$
15. $f(x) = 3x^4 + 2x^2 - 8$
16. $y = 2$
17. $y = x$
18. $f(x) = x^3 - 1$
19. $f(x) = x^2 - 1$
20. $f(x) = -x + \sin x$
21. $f(x) = -x$
22. $f(x) = x^2$

23. $f(x) = \dfrac{1}{x^2}$

24. $f(x) = 2x^4$

25. $f(x) = x^3 + 1$

26. $f(x) = \dfrac{x}{x-2}$

27. $f(x)\ x^3 + x$

28. $f(x) = \sin(x)$

29. $f(x) = \sqrt{x^2 + 1}$

30. $f(x) = \cos x$

    $g(x) = 2x+1$

    ii) $f(g(x))$

    iii) $g(f(x))$

4.13 Graphs of Trigonometric Functions

1. As x increases from 0 to π, what happens to $2\sin\dfrac{x}{2}$?

2. What is the amplitude, Axis of wave and offset of $y=5\sin(x)+12\cos(x)-2$?

3. What is the maximum value of $y = \sqrt{4 + \cos^2 x}$ in the interval $\left[\dfrac{-\pi}{2}, \dfrac{\pi}{2}\right]$

4. Find y intercept of the function $y = \left|\sqrt{3}\sec\left[3(x+\dfrac{\pi}{4})\right]\right|$

5. Find amplitude of the function $f(x) = -\dfrac{1}{2}\sin(x)\cos(x) + 1$

6. Find the primary period of $f(x) = \dfrac{\cos(2x)}{1+\sin(2x)}$

7. Find primary period of $f(x)=3\sin^2(2x)$

8. Find y intercept of $y = \sqrt{3}\sin(x+\dfrac{\pi}{3})$

9. What is the amplitude of the function $y=3\sin x+4\cos x+1$

10. Find maximum value of the function $f(x) = \sin(\dfrac{x}{4})$ over the interval $0 \le x \le \dfrac{\pi}{3}$

11. Find maximum value of $4 \sin x \cos x$

12. What happens to $\sin x$ as x increases from $-\dfrac{\pi}{4}$ to $\dfrac{3\pi}{4}$?

13. What is the smallest positive x intercept of $y = 2\sin\left[3(x+\frac{3\pi}{4})\right]$?

14. What is the smallest positive angle that will make $y = 3+\sin\left[3(x+\frac{\pi}{3})\right]$ a minimum?

15. Find amplitude of the graph of the function $y=\cos^4 x-\sin^4 x+1$

16. Find amplitude, period and frequency of the following:
    1. $y=2\sin(\pi x+\pi)$
    2. $y = \frac{3}{4}\cos(\frac{x}{2}-\frac{\pi}{2})$

17. Find the coordinates of the first maximum point in the graph of $y = \sin(\frac{x}{2})$ that has a positive x-coordinate.

4.14 Miscellaneous Graphs

1. Find the point of intersection of the graphs $y=\log x$ and $y=\ln\frac{x}{2}$.

2. At how many points does the function $y=x^3+5x-2$ intersect the x axis?

3. Plot the graph of $f(x)=\frac{x^2-1}{x-1}$; locate the hole that the function has.

4. Find x and y intercept(s) of the graph of equation $y=(x^2-4)\ln(x^2+9)$

5. Determine which of the following functions has an inverse that is also a function.
   a. $y = x^2 - 3x + 5$ b. $y = |x+2| - 1$
   c. $y = \sqrt{16-9x^2}$ d. $y=x^3+5x-2$

6. $f(x)=2x^2+12x+3$; if the graph of $f(x-k)$ is symmetric about the y axis, what is k?

7. Find equation of the axis of symmetry of $y=3x^2-x+2$.

8. Find the distance between the x and y intercepts of the function $f(x)=2x^3+x+1$.

9. $y=-2x^2+4x-7$

    Determine the coordinates of the vertex of the parabola given above. Does the above function have a maximum or minimum? What is this value? Find the equation of the axis of symmetry also.

10. Plot the graphs of the following:
    a. $\frac{x^2}{9}+\frac{y^2}{4}=1$ b. $\frac{x^2}{4}+\frac{y^2}{9}=1$ c. $\frac{x^2}{9}-\frac{y^2}{4}=1$ d. $\frac{y^2}{9}-\frac{x^2}{4}=1$

Advanced Calculation and Graphing Techniques with the TI – 83 Plus Graphing Calculator

Unauthorized copying or reuse of any part of this page is illegal.

4.15 The Greatest Integer Function

1. What is the period and frequency of the function $f(x)=|1-2x+2[x]|$ if $[x]$ represents the greatest integer less than or equal to x? What are the maximum and the minimum values of f(x)? What is the amplitude, offset, and equation of the Axis of wave? What is the domain and range?

2. $g(x)=[x]-2x+1$ what is the period of g(x)?

3. $f(x)=k$ where k is an integer for which $k \leq x < k+1$ and $g(x) = |f(x)|-f(x)+1$. What is the minimum value for g(x)?

4. $f(x)=[x]$ where $[x]$ represents the greatest integer function. What is the range of f(x)?

5. $[4.6]- [-5.4]+2[0.3]+ [4]- [0]=?$

4.16 Parametric Graphing

Plot the graph and state what each curve represents.

1. $x = 4\cos\theta+1$
   $y = 3\sin\theta-1$

2. $x = t^2+t+1$
   $y = t^2-t+1$

3. $x = t^3+2$
   $y = \frac{4}{3}t^3+1$

4. $x = t^2$
   $y = 2t^2-1$

5. $x = \sin\theta$
   $y = \cos\theta$

6. $x = t$
   $y = \sqrt{4-t^2}$

7. $x = \sqrt{p}$
   $y = \sqrt{4-p}$

8. $x = 2\sin\alpha$
   $y = 2\sin\alpha$

9. $x = 2\sin\alpha$
   $y = 2\cos\alpha$

10. $x = 3\sin\alpha$

y = 4cosα

11. x = 3t+4

    y = t−6

12. x = sin²t

    y = 3cost

13. x = t(1+t)

    y = t(−1+t)

14. Using parametric equations plot the set of points (x²,y) where y=x²+1

15. Using parametric equations plot the set of points (x²,y) where y=2x−1.

16. Plot the inverse of the following function using parametric equations. y=2x³+x+1

4.17 Polar Graphing

1. What is the area enclosed by the following curves and the coordinate axes?

    $r = \dfrac{3}{\sin\theta}$

    $r = \dfrac{4}{\cos\theta}$

2. $r = \dfrac{4}{\dfrac{1}{\sec\theta} + 2\sin\theta}$ what is the area of the region bounded by the above curve and the x and y axes?

3. What is the area of the region that the curve r=3cosθ represents?

4.18 Limits

1. $\lim\limits_{x\to\infty} \dfrac{3x^4 - 5x^3 + 8}{-4x^4 + 7x^2 + 4x + 5} = ?$

2. $\lim\limits_{x\to\infty} \dfrac{6x^3 + 5x^2 - 8x}{-2x^2 + 1} = ?$

3. $\lim\limits_{x\to 2} \dfrac{-x^2 + 4}{x^3 + 8} = ?$

4. $\lim\limits_{x\to -\infty} \dfrac{x^3 - 27}{x^4 - 81} = ?$

5. $\lim\limits_{x\to 3} \dfrac{x^3 - 27}{x^4 - 81} = ?$

6. $\lim\limits_{x\to\infty} \dfrac{6x^3 - 9x + 1}{5x^3 - 7} = ?$

Advanced Calculation and Graphing Techniques with the TI – 83 Plus Graphing Calculator

7. $\lim_{x \to 2^+} \dfrac{3x+5}{x-2} = ?$

8. $\lim_{x \to 2} \dfrac{3x+5}{x-2} = ?$

4.19 Continuity

1. $f(x) = \begin{cases} \dfrac{4x^2+3x}{x} & x \neq 0 \\ m & x = 0 \end{cases}$

    m=? if f(x) is a continuous function.

2. In order to be continuous at x=2 what must $f(x) = \dfrac{x^4-16}{x^3-8}$ be defined to be equal to?

3. $f(x) = \begin{cases} \dfrac{6x^2-6}{x-1} & x \neq 1 \\ A & x = 1 \end{cases}$

    What must be A if f(x) is a continuous function?

4.20 Horizontal and Vertical Asymptotes

1. Find the horizontal and vertical asymptotes as well as the domain and range of:

    a) $y = \dfrac{2x^2-18}{x^2-4}$

    b) $y = \dfrac{x+2}{x^2-4}$

    c) $y = \dfrac{x^2-4x-5}{x^2-1}$

    d) $y = \dfrac{x+3}{(x-3)(x^2-9)}$

2. Find equations of the vertical asymptotes of $f(x) = \dfrac{x^2+4x+3}{x+2} \cdot \dfrac{1}{\cot(\pi x)}$

4.21 Complex Numbers

1. $4(\text{cis}\,70°)^4 = ?$
2. $f(x) = 3x^5 - 2x^3 + 8x - 2 \Rightarrow f(i) = ?$
3. If n is an arbitrary positive integer then $i^{4n+5} + i^{4n+6} + i^{4n+7} + i^{4n+8} = ?$
4. $i^{192} + i^{193} + i^{194} + i^{195} = ?$
5. What is the reciprocal of $3+4i$

Advanced Calculation and Graphing Techniques with the TI – 83 Plus Graphing Calculator

6. $\dfrac{1+i}{6i+8} = ?$

7. $z = 3\operatorname{cis}\dfrac{\pi}{8} \Rightarrow z^3 = ?$

8. $z = \dfrac{1+i\sqrt{3}}{-1+i\sqrt{3}}$, what is the value of z in trigonometric form?

9. $A = 3\operatorname{cis}40°$
   $B = 4(\cos 50° + i\sin 50°)$
   $A \cdot B = ?$

4.22 Permutations and Combinations

1. $_5P_2 + {}^6P_3 + P(5,3) = ?$

2. $\dbinom{5}{3} + C_2^8 + {}_6C_3 = ?$

3. $\dfrac{(6+3)!}{6! + 3!}$

4.23 Miscellaneous Calculations

1. $\cos(2\sin^{-1}(\dfrac{-5}{13})) = ?$

2. $f(x) = \sin(\cos(x))$
   $f(320°) = ?$

3. $f(x) = 14x^2$
   $g(x) = f(\cos x) + f(\tan x)$
   $g(12°) = ?$

4. $\cos 30 \cdot \cos 30° - \sin 45 \cdot \sin 45° = ?$

5. $f(r, \theta) = r\cos\theta$
   $f(12, 13) = ?$

6. $\csc(\tan^{-1}\dfrac{1}{\sqrt{2}}) = ?$

7. $\operatorname{Arcsin} 0.6 + \cos^{-1} 0.6 = ?$

8. $\cos x = \dfrac{1}{3} \Rightarrow \sqrt{\sec x} = ?$

9. $\left.\begin{array}{l} f(x,y) = \tan(x) + \tan(y) \\ g(x,y) = 1 - \tan(x)\tan(y) \end{array}\right\} \dfrac{f(10°, 20°)}{g(10°, 20°)} = ?$

10. $\cos A = 0.6631$

tanA= -1.12884

$\Rightarrow$ A=? (in degrees)

11. sec 4.1= x

   csc (3 Arctanx)=?

12. sin A= $\frac{5}{13}$ 90° ≤ A ≤ 180°

   cosB= $\frac{4}{9}$, B is in 4'th quadrant.

   sin (A+B)= ?

13. In which quadrant is the angle represented by Arcsin ($\frac{-3}{5}$) + Arccos ($\frac{-12}{13}$)?

14. A # B = $\sqrt{cosA + secB}$

   4 # 5= ?

15. $\frac{\cos 15°}{\sin 75°}$ =?

16. $\frac{\tan 25°}{\cos 65°}$ = ?

17. cos (tan$^{-1}\frac{1}{3}$)= ?

18. $\frac{\sin 135° \cdot \cos \frac{5\pi}{6}}{\tan 225°}$ =?

19. cscθ= $\frac{4}{3}$

   cosθ < 0

   tanθ= ?

20. cos310° + sin140°= 2x

   $\Rightarrow$ x = ?

21. cos π - sin 930° – csc $\left(\frac{-5\pi}{2}\right)$ + sec (0°)= ?

22. tan (-135°) + cot $\left(\frac{-7\pi}{8}\right)$ = ?

23. tan (-30°)= - cot x $\Rightarrow$ x= ? (x is in 3rd quadrant)

24. cos 210°= ?

Advanced Calculation and Graphing Techniques with the TI – 83 Plus Graphing Calculator

25. $\sin 330° = ?$

26. $\sec \dfrac{7\pi}{6} \cdot \tan \dfrac{3\pi}{4} \cdot \sin \dfrac{2\pi}{3} = ?$

27. A is in 3rd quadrant and $\tan A = \dfrac{8}{15}$

    B is in 2nd quadrant and $\tan B = \dfrac{-3}{4}$

    In which quadrant does (A+B) lie?

28. Given that $\tan\theta = \dfrac{-5}{12}$ and $\sin\theta$ is positive.

    $\cos(2\theta) - \sin(180° - \theta) = ?$

29. A and B are acute angles and

    $\tan(A) = \dfrac{12}{5}$ and $\sin(B) = \dfrac{4}{5} \Rightarrow \cos(2A + B) = ?$

30. $\cos\theta = \dfrac{4}{5}$

    $\sin(2\theta) = ?$

    $\tan(2\theta) = ?$

31. $\dfrac{\tan 100° + \tan 35°}{1 - \tan 100° \cdot \tan 35°} = ?$

32. $\tan\theta = \dfrac{3}{4} \Rightarrow \sin\theta = ?$

33. $\cos^2 20° - \sin^2 20° = ?$

34. $\sin A = \dfrac{3}{5}$ and $\cos A < 0$

    $\tan(2A) = ?$

35. Important Note: The answers for the following questions are given in both Degree and Radian values as both of them may be asked in the test. It should be noted, however, that these are not two separate answers, but different representations of the same correct answer.

    a. $\sin^{-1} \dfrac{1}{2} = ?$

    b. $\sin^{-1} \dfrac{-\sqrt{3}}{2} = ?$

c. $\cos^{-1}\dfrac{\sqrt{3}}{2}=?$

d. $\text{Arccos}\dfrac{-\sqrt{3}}{2}=?$

e. $\tan^{-1}(1)=?$

f. $\text{Arctan}(-1)=?$

g. $\cot^{-1}(\dfrac{1}{\sqrt{3}})=?$

h. $\cot^{-1}(\dfrac{-1}{\sqrt{3}})=?$

i. $\sec^{-1}(\sqrt{2})=?$

j. $\sec^{-1}(-\sqrt{2})=?$

k. $\csc^{-1}(2)=?$

l. $\csc^{-1}(-2)=?$

36. $\sin(\text{Arccos}\dfrac{4}{5})=?$

37. $\cos(\text{Arcsin}\dfrac{-4}{5}+\text{Arccos}\dfrac{12}{13})=?$

38. $\cos(\sin^{-1}(\dfrac{-1}{2}))=?$

39. $f(x)=\sin x$

    $f^{-1}(\dfrac{3\pi}{14})=?$

40. $\sin(2\arctan(\dfrac{-15}{8}))=?$

41. $\tan(\text{Arccos}\dfrac{-3}{5})=?$

42. a. $\text{Arccos}(\cos\dfrac{7\pi}{6})=?$

    b. $\text{Arctan}(\tan\dfrac{\pi}{4})=?$

    c. $\text{Arctan}(\tan\dfrac{5\pi}{4})=?$

43. $f(x)=e^x$

    $g(x)=\cos x$

(fog)($\sqrt{3}$)= ?

44. f(x)= $\sqrt{2x-2}$

    g(x)= cosx

    g^{-1} (f($\sqrt{2}$))=?

45. $\sqrt{2002.2003 - 2001.2002}$ = ?

46. f(x,y)= $2x^2 - y^2$

    g(x)= 5^x

    g(f(4,3))= ?

47. f(x,y)= $\sqrt{3x^2 - 4y}$

    g(x)= 3^x

    g(f(2,1)) = ?

48. f(x)= $\sqrt{3x-4}$

    g(x)= $x^3 + x + 1$

    f(g(2))= ?

49. f(x)=3x

    f($\log_7 4$)=?

50. f(x)= $x^2 + 5x - 7$

    g(x)= x – 5

    f(g(ln9))= ?

51. f(x)= $\sqrt{x}$

    g(x)= $\sqrt[3]{(x+2)^2}$

    h(x)= $\sqrt[5]{x-4}$

    h(g(f(2)))= ?

52. f(x)= x lnx

    g(x)= 10^{x+1}

    g(f(3))=?

53. $\sin^{-1}(\cos 200°)$=?

54. $\sqrt[3]{y} = 2.6$

    $\sqrt[4]{10y} = ?$

55. $a\Omega b = \dfrac{a}{e + \dfrac{\pi}{b}}$

(2 Ω 3) Ω 4 = ?

56. $\left(\dfrac{28}{34}\right)^{\frac{-5}{6}} = ?$

57. $\sqrt{3}\cdot\sqrt[3]{4}\cdot\sqrt[4]{5} = ?$

58. $a \diamond b = \dfrac{\sqrt[3]{a}+\sqrt[3]{2b}-1}{\sqrt{ab+1}}$

    $3 \diamond \pi = ?$

59. $\left(-\dfrac{2}{9}\right)^{3/5} = ?$

60. $f(x) = 3x^2 + 8x - 6$

    Find the negative value of $f^{-1}(0)$

61. $f(x) = x\sqrt[3]{x}$

    $(f(\sqrt{2}) = ?$

62. $x_o = \sqrt{2}$

    $x_{n+1} = x_n \sqrt[3]{x_n + 2}$

    $x_4 = ?$

63. $f(x) = \sqrt[3]{x}$

    $g(x) = x^4 + 2$

    $(f \circ g)(4) = ?$

64. $3^{4/3} + 4^{5/4} = ?$

65. $f(x) = |x| + [x]$

    $f(1.5) - f(-4.5) = ?$

66. $\log_4(\cos 290°) = ?$

67. $\sum_{i=9}^{12} \ln i = ?$

68. $\log_6 3 = ?$

69. $\log(\sin 2) + \log(\sin 20) + \log(\sin 20°) = ?$

70. $F(x,y) = \log_y x$

    $F(e, \pi^2) = ?$

71. $\log_{36} 6 - \log_3 27 + \log_2 (0.25)^{1/3} = ?$

72. $\log_{\sqrt{5}} 4 - \log_{16} \sqrt{125} = ?$

73. $f(x) = x^4 - 88x^3 - 1134x^2 + 3888x + 56135$

    $f(99) = ?$

74. $f(x) = x^3 - 4x^2 + 6x - 4$

    $f(3) - f(\sqrt{3}) = ?$

75. What is the remainder when

    $3x^4 + 8x^3 + 9x^2 - 3x - 4$ is divided by $x + 1$?

NOTES

CHAPTER 5.
SOLUTIONS TO THE SAMPLE TI PROBLEMS

NOTES

Advanced Calculation and Graphing Techniques with the TI – 83 Plus Graphing Calculator

5.1 Polynomial Equations

When solving a polynomial or algebraic equation in the form **f(x)=g(x)**, perform the following steps:

i. Write the equation in the form: **f(x)-g(x)=0**.

ii. Plot the graph of **y=f(x)-g(x)**.

iii. Find the x-intercepts using the **Calc Zero** of TI-83 Plus. However when the graph seems to be tangent to the x-axis at a certain point, you may use the **Calc Min** or **Calc Max** facilities but you should make sure that the y-coordinate of the minimum or maximum point is zero.

iv. Any value like **-6.61E -10** or **7.2E -11** can be interpreted as 0 as they mean **-6.6x10^{-10}** and **7.2x10^{-11}** respectively.

1. $P(x) = 3x^3 - 5x^2 + 6x - 3$

   The zero of the above polynomial lies between two consecutive integers. What are these integers?

   Solution:

   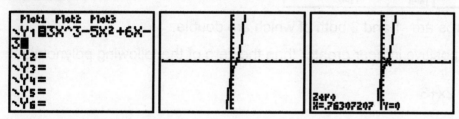

   Answer: Between 0 and 1

2. $y = 3x^2 - 4x - 5$

   What is the positive zero of the above function correct to the nearest hundredth?

   Solution:

   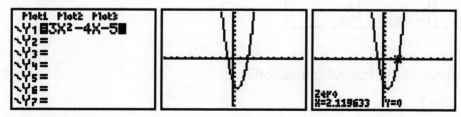

   Answer: 2.12

3. Is x-99 a factor of the following polynomial?

   $P(x) = 2x^4 - 200x^3 + 194x^2 + 400x - 394$

   Solution:

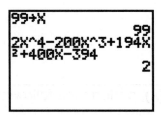

Answer: NO because P(99) is not zero.

4. Find all real zeros of the following polynomial:

$P(x) = 2x^6 - 2x^5 - 8x^4 - 2x^3 + 10x^2 + 16x + 8$

Solution:

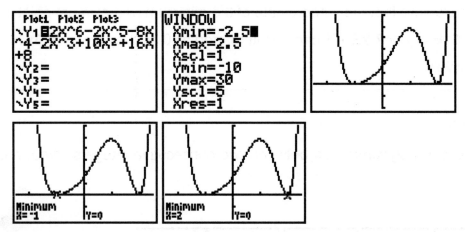

Answer: Real zeros are -1 and 2 both of which are double.

5. What is the least positive integer greater than the zero of the following polynomial?

$P(x) = -\dfrac{3}{2}x^3 - x^2 - 2x + 3$

Solution:

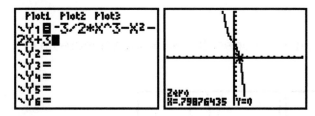

Answer: 1

6. What are the real roots of the following function?

$f(x) = -2x^4 - 4x^3 + 6x^2 - 4x + 7$

Solution:

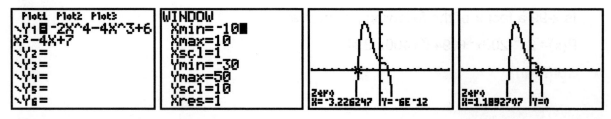

Answer: -3.226 and 1.189

7. $P(x) = 3x^4 - x^3 + 2x^2 + 5x - 1$

   How many positive and negative real zeros does the above polynomial have?

   Solution:

   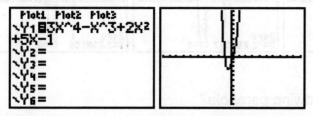

   Answer: 1 positive and 1 negative real zero.

8. $x^2 + x + 2 = 0$

   What is the nature of the roots of the above equation?

   Solution:

   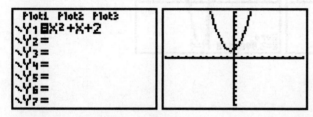

   Answer: No real zeros, two complex conjugate zeros.

9. $P(x) = 2x^3 + x^2 + 3x - 5$

   How many positive and negative real zeros does the above polynomial have?

   Solution:

   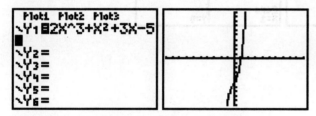

   Answer: 1 positive real zero only.

10. Find the positive rational root of the following equation:

    $2x^3 - 5x^2 + 14x = 35$

    Solution:

    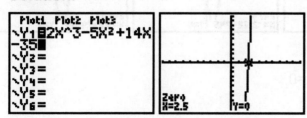

    Answer: 2.5 or 5/2

11. What is the absolute difference between the zeros of the following function?

    $f(x) = 7x^2 + 11.5x - 25$

Solution:

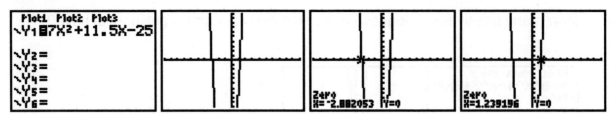

Answer: 1.239−(−2.882)=4.121

12. What is the sum of the zeros of the following parabola?

    y = $3x^2 - 7x - 5$

    Solution:

    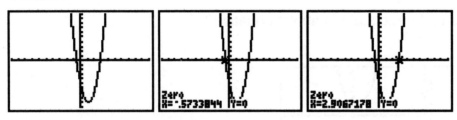

    Answer: 2.907+(−0.573)=2.33

13. What are the zeros of y = $3x^2 + x - 4$?

    Solution:

    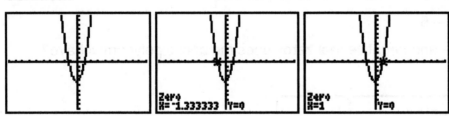

    Answer: 1 and −1.333

14. f(x) = $6x^2 + 12x - 3$

    f(q)=0 ⇒ What is one value of q?

    Solution:

    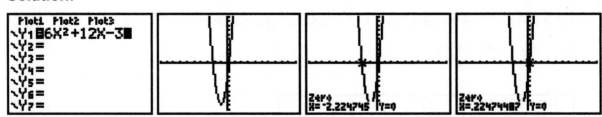

    Answer: −2.225 or 0.225.

15. Find the sum of the roots of $6x^3 + 8x^2 - 8x = 0$

    Solution:

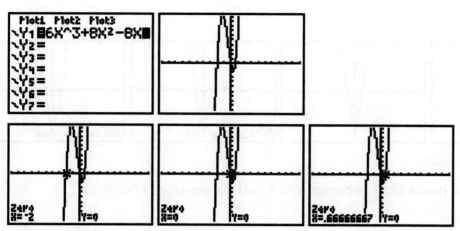

Answer: -2 + 0 + .667 = -1.333

16. $P(x) = 2x^2 + 3x + 1$

   $P(a) = 7 \Rightarrow a = ?$

   Solution:

   $2a^2 + 3a + 1 = 7$

   $2a^2 + 3a + 1 - 7 = 0$

   $2a^2 + 3a - 6 = 0$

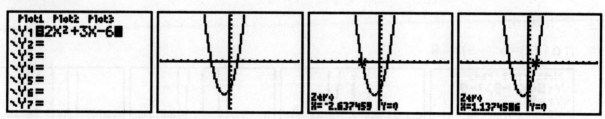

Answer: -2.637 or 1.137

17. What is the product of the roots of the following equation $(x - \sqrt{3})(x^2 - ex - \pi) = 0$?

   Solution:

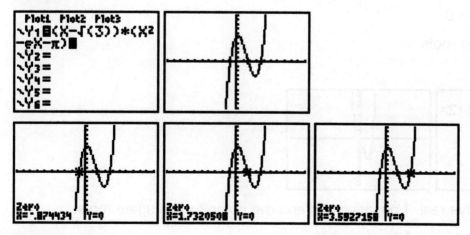

Answer: -5.44

18. $f(x) = 5x^2 - 7$

   Find sum of the zeros of $f(x)$.

Solution:

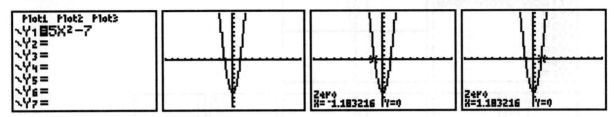

Answer: -1.18+1.18=0

19. $P(x) = x^3 + 6x - 14$ has a zero between which two consecutive integers?

   Solution:

   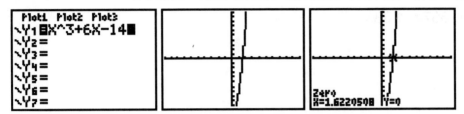

   Answer: Between 1 and 2.

20. $f(x) = x^2 - 9$

   $(f \circ f)(x) = 0 \Rightarrow$ What are the real values of x?

   Solution:

   $(f \circ f)(x) = (x^2 - 9)^2 - 9$

   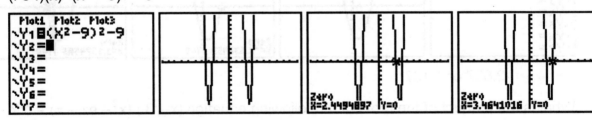

   Answer: -2.449, 2.449, -3.464 or 3.464

21. $2x^4 + 3x^3 + 2x - 1 = 0$

   Find nature of the roots.

   Solution:

   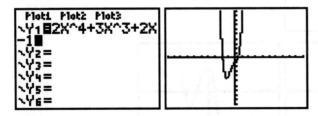

   Answer: 1 positive real, 1 negative real and two complex conjugate roots.

22. Is $3x+1$ a factor of $2x^3 + 4x^2 - 4x - 3$?

   Solution:

   If $3x+1$ is a factor then $-1/3$ must be a zero of the polynomial.

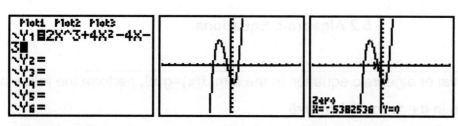

Answer: No

23. Find the number of the positive real zeros of the following equation:

    $x^4 + 2x^3 - 4x^2 - 5x = 0$

    Solution:

    $x^4 + 2x^3 - 4x^2 - 5x = 0$

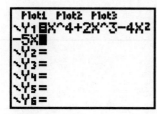
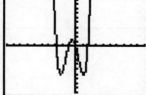
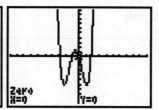

Answer: 1

24. Find product of the real roots of the following equation:

    $x^4 - 3x^3 - 72x^2 - 3x - 18 = 0$

    Solution:

    $x^4 - 3x^3 - 72x^2 - 3x - 18 = 0$

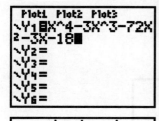
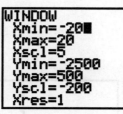
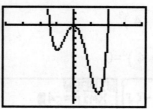

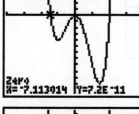
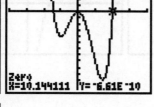

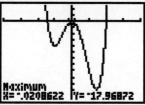

Answer: $-7.11 \times 10.14 = -72.09$

Advanced Calculation and Graphing Techniques with the TI – 83 Plus Graphing Calculator

5.2 Algebraic Equations

When solving a polynomial or algebraic equation in the form **f(x)=g(x),** perform the following steps:
i.   Write the equation in the form: **f(x)-g(x)=0**.
ii. Plot the graph of **y=f(x)-g(x)**.
iii. Find the x-intercepts using the **Calc Zero** of TI-83 Plus. However when the graph seems to be tangent to the x-axis at a certain point, you may use the **Calc Min** or **Calc Max** facilities but you should make sure that the y-coordinate of the minimum or maximum point is zero.
iv. Any value like **-6.61E -10** or **7.2E -11** can be interpreted as 0 as they mean **-6.6x10^{-10}** and **7.2x10^{-11}** respectively.

1. $f(x) = \sqrt{3x+4}$

   $g(x) = x^3$

   If is given what (fog)(x)=(gof)(x), then what is x?

   Solution:

   $(fog)(x) = \sqrt{3x^3+4}$

   $(gof)(x) = (\sqrt{3x+4})^3$

   $\sqrt{3x^3+4} = (\sqrt{3x+4})^3$

   $\sqrt{3x^3+4} - (\sqrt{3x+4})^3 = 0$

     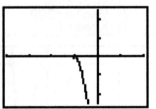

   Answer: -1

2. $f(x) = \sqrt{-x^3+4x}$

   $g(x) = 4x$

   What is the sum of the roots of the equation f(x)=g(x)?

   Solution:

   $\sqrt{-x^3+4x} = 4x$

   $\sqrt{-x^3+4x} - 4x = 0$

Advanced Calculation and Graphing Techniques with the TI – 83 Plus Graphing Calculator

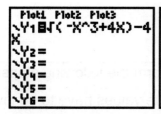 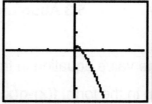 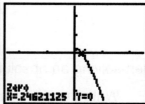

Answer: 0.246

3. $a \# b = a^b - b^a$

   If $3 \# k = k \# 2$ then $k=?$

   Solution:

   $3^k - k^3 = k^2 - 2^k$

   $3^k - k^3 - k^2 + 2^k = 0$

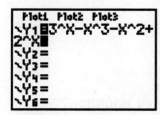 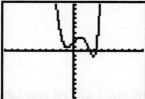 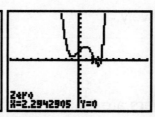 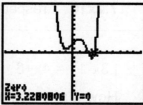

Answer: 2.294 or 3.228

4. $5x^{4/3} = 2 \Rightarrow x=?$

   Solution:

   $5x^{4/3} = 2 \Rightarrow 5x^{4/3} - 2 = 0$

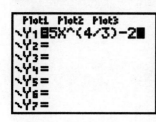 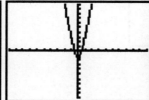 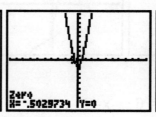 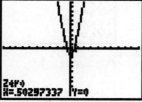

Answer: 0.503 or -0.503.

5. Find sum of the roots of: $2x - \dfrac{5}{x} + 2 = 0$

   Solution:

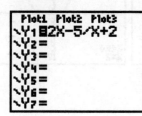 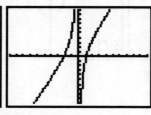 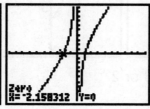 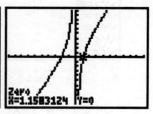

Answer: -2.158 + 1.158 = -1

Advanced Calculation and Graphing Techniques with the TI – 83 Plus Graphing Calculator

5.3 Absolute Value Equations

When solving an absolute value equation in the form **f(x)=g(x),** perform the following steps:

i. Write the equation in the form: **f(x)-g(x)=0**. Whenever absolute values have to be involved, replace **| f(x) |** by **abs(f(x))**

ii. Plot the graph of **y=f(x)-g(x)**.

iii. Find the x-intercepts using the **Calc Zero** of TI-83 Plus. However when the graph seems to be tangent to the x-axis at a certain point, you may use the **Calc Min** or **Calc Max** facilities but you should make sure that the y-coordinate of the minimum or maximum point is zero.

iv. Any value like **-6.61E -10** or **7.2E -11** can be interpreted as 0 as they mean **-6.6x10^{-10}** and **7.2x10^{-11}** respectively.

1. | 3x-1 | = 4x+6

   How many numbers are there in the solution set of the above equation?

   Solution:

   | 3x-1 | - (4x+6)=0

   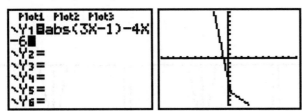

   Answer: 1

2. | x-3 | + | 2x+1 | = 6 ⇒ x=?

   Solution:

   | x-3 | + | 2x+1 | - 6 = 0

   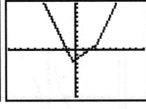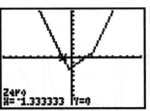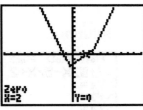

   Answer: x= -1.33 or 2

3. | 3x-5 | = 4 ⇒ x=?

   Solution:

   | 3x-5 | - 4 = 0

Advanced Calculation and Graphing Techniques with the TI – 83 Plus Graphing Calculator

 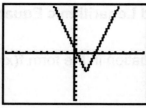 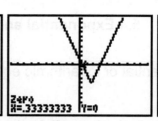 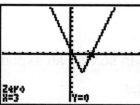

Answer: x=0.33 or 3

4. $|4x+6| = 3x+4 \Rightarrow x=?$

Solution:

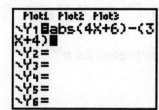 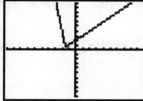 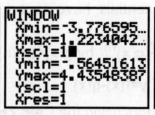 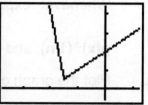

Answer: No solution

5. $\dfrac{|x-3|}{x} = 4 \Rightarrow x = ?$

Solution:

$\dfrac{|x-3|}{x} - 4 = 0$

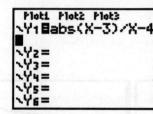 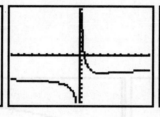 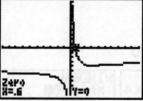

Answer: 0.6

Advanced Calculation and Graphing Techniques with the TI – 83 Plus Graphing Calculator

5.4 Exponential and Logarithmic Equations

When solving an exponential or logarithmic equation in the form **f(x)=g(x)**, perform the following steps:

i. Write the equation in the form: **f(x)-g(x)=0**.

ii. Whenever logarithms have to be involved, replace **log $_{f(x)}$ g(x)** by $\dfrac{\log g(x)}{\log f(x)}$ or by $\dfrac{\ln g(x)}{\ln f(x)}$.

Whenever exponentials have to be involved, replace **f(x) $^{g(x)}$** by **f(x)^g(x)**; $\sqrt[n]{g(x)}$ by **g(x)^(1/n)**; and $\sqrt[n]{g(x)^m}$ by **g(x)^(m/n)**; **exp(f(x))** must be interpreted as **e$^{f(x)}$**.

iii. Plot the graph of **y=f(x)-g(x)**.

iv. Find the x-intercepts using the **Calc Zero** of TI-83 Plus. However when the graph seems to be tangent to the x-axis at a certain point, you may use the **Calc Min** or **Calc Max** facilities but you should make sure that the y-coordinate of the minimum or maximum point is zero.

v. Any value like **-6.61E -10** or **7.2E -11** can be interpreted as 0 as they mean **-6.6x10 $^{-10}$** and **7.2x10 $^{-11}$** respectively.

---

1. $\log_4 x \cdot \log_5 6 = 7 \Rightarrow x = ?$

    Solution:

    $\log_4 x \cdot \log_5 6 - 7 = 0$

    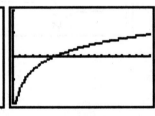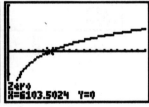

    Answer: 6103.5

2. $A = e^{Bt}$

    A=1000, T=4, B=?

    Solution:

    $1000 = e^{B*4}$

    $1000 - e^{B*4} = 0$

    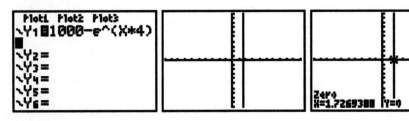

Answer: 1.73

3. $f(x) = \exp(x)$

   $(\exp(x) = e^x)$

   If $h(x) = f(-x) + f^{-1}(-x)$ then $h(-2) = ?$

   Solution:

   $h(-2) = f(2) + f^{-1}(2)$

   $f(2) = e^2 = 7.389$

   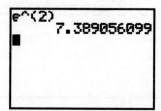

   $f^{-1}(2) = x \Rightarrow f(x) = e^x = 2 \Rightarrow e^x - 2 = 0$

   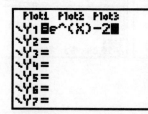 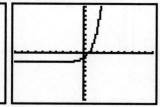 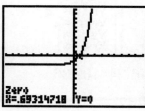

   Answer: $7.389 + 0.693 = 8.082$

4. If $\log x = \dfrac{3}{4}$ then $\log(1000x^2) = ?$

   Solution:

   $\log x - \dfrac{3}{4} = 0$

   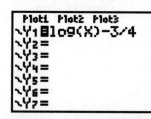 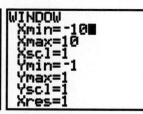 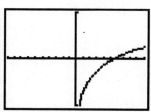 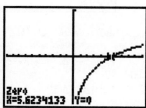

   Answer: 4.5

5. $\left.\begin{array}{l}\log_3 x = \sqrt{5} \\ \log_5 y = \sqrt{3}\end{array}\right\}$ $x \cdot y = ?$

Solution:

$\log_3 x = \sqrt{5}$

$\log_3 x - \sqrt{5} = 0$

 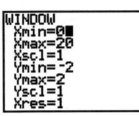 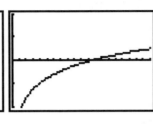 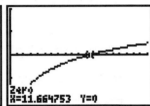

$\log_5 y = \sqrt{3}$

$\log_5 y - \sqrt{3} = 0$

 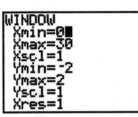 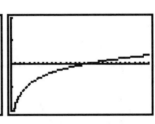 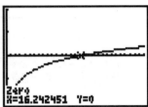

Answer: 11.66 * 16.24 = 189.36

6. $\log_3 2 = x \cdot \log_6 5 \Rightarrow x = ?$

   Solution:

   $\log_3 2 - x \cdot \log_6 5 = 0$

    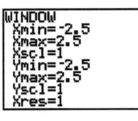 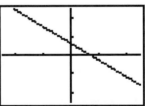 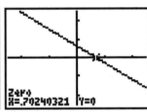

   Answer: 0.702

7. $2^{x+3} = 3^x \Rightarrow x = ?$

   Solution:

   $2^{x+3} - 3^x = 0$

    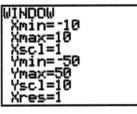 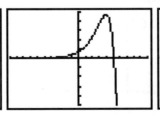 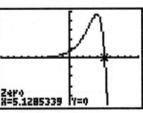

   Answer: 5.129

8. $\log_x 3 = \log_4 x \Rightarrow$ What is the sum of the roots of this equation?

   Solution:

$\log_x 3 - \log_4 x = 0$

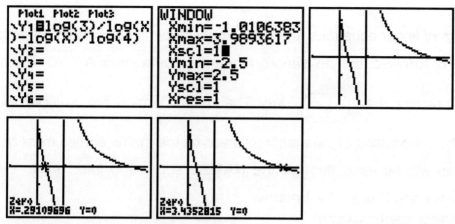

Answer: 0.291+3.435=3.726

9. $f(x)=3.5^x+1$; $f^{-1}(10)=?$

Solution:

$f^{-1}(10)=x \Rightarrow f(x)=10$

$3.5^x+1=10$

$3.5^x - 9=0$

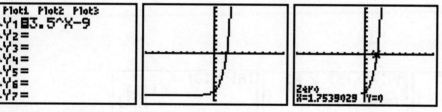

Answer: 1.754

10. $3.281^x = 4.789^y \Rightarrow \dfrac{x}{y} = ?$

Solution:

If $y=1$ then $\dfrac{x}{y}=x \Rightarrow 3.281^x = 4.789^1 = 4.789$

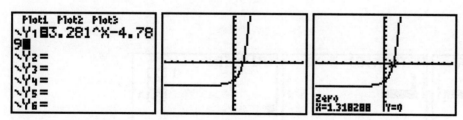

Answer: 1.318

Advanced Calculation and Graphing Techniques with the TI – 83 Plus Graphing Calculator

5.5 System of Linear Equations, Matrices and Determinants

In order to solve a system of linear equations, the coefficients of the linear system are entered in two separate matrices A and B followed by the simple algebraic operation of $A^{-1} * B$. The matrix menu is accessed by pressing the key. The matrix entries are entered through the **EDIT** sub menu where the dimensions of the matrix followed by the matrix entries must be input. The matrices that are entered will be used through the **NAMES** sub menu later on. In the following example A is a 2 by 2 matrix and B is a 2 by 1 matrix.

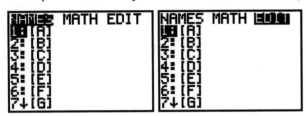

1. $\begin{matrix} x + 3y = 7 \\ 12x - 2y = 8 \end{matrix} \Big\} \dfrac{x}{y} = ?$

   Solution:

   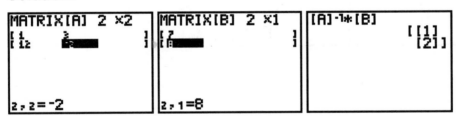

   Answer: 1/ 2

2. $\begin{matrix} x + y + z = 6 \\ 2x - y + 3z = 9 \\ 3x + y - 4z = -7 \end{matrix} \Big\} x^2+y^2+z^2=?$

   Solution:

   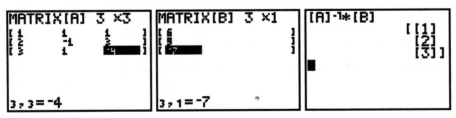

   Answer: $1^2+2^2+3^2=1+4+9=14$

3. If $A=\begin{bmatrix} 2 & 3 \\ -1 & 5 \end{bmatrix}$, $B=\begin{bmatrix} 0 & -2 \\ -3 & 5 \end{bmatrix}$, Find (i) A+ B; (ii) 3A – 2B; (iii) AB; (iv) BA

Solution:

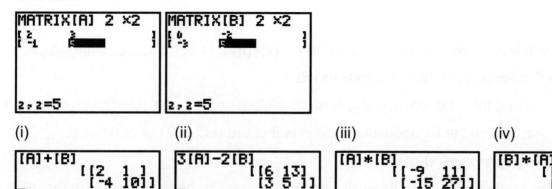

(i)   (ii)   (iii)   (iv)

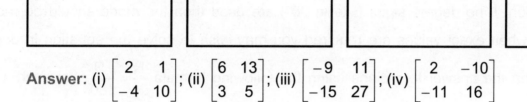

Answer: (i) $\begin{bmatrix} 2 & 1 \\ -4 & 10 \end{bmatrix}$; (ii) $\begin{bmatrix} 6 & 13 \\ 3 & 5 \end{bmatrix}$; (iii) $\begin{bmatrix} -9 & 11 \\ -15 & 27 \end{bmatrix}$; (iv) $\begin{bmatrix} 2 & -10 \\ -11 & 16 \end{bmatrix}$

4. Find the determinant and inverse of the matrix $\begin{bmatrix} 1 & 3 & -1 \\ -2 & 4 & 1 \\ 0 & 0 & 2 \end{bmatrix}$

Solution:

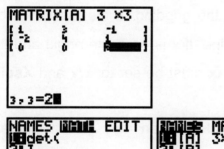

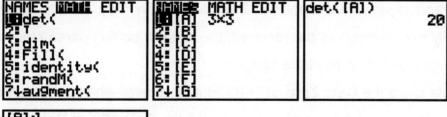

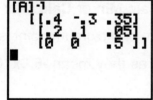

Answer: $A^{-1} = \begin{bmatrix} 0.4 & -0.3 & 0.35 \\ 0.2 & 0.1 & 0.05 \\ 0 & 0 & 0.5 \end{bmatrix}$

5.6 Trigonometric Equations

When solving a trigonometric equation in the form **f(x)=g(x),** perform the following steps:

i. Write the equation in the form: **f(x)-g(x)=0**.

ii. While writing the trigonometric expressions please observe the rules given in parts 1.17 and 1.18 that involve the trigonometric functions that are built in TI or otherwise.

iii. Plot the graph of **y=f(x)-g(x)**.

iv. Set the **angle mode to radians or degrees** depending on which angle measure is used in the question. If no degree signs (like in 90°) are used then the mode should be radians. However when exact values are required you may wish to solve the equation in degrees and convert the answer to radians using the following formula $\frac{R}{\pi} = \frac{D}{180°}$. In such a case finding the answer in radians and then trying to find which answer choice matches this answer can also be an option; while doing so you may directly replace π with 180°

v. If x is limited to a certain interval then set **Xmin**; **Xmax** and **Xscl** accordingly. For example, if x is an acute angle and the angle mode is degrees, then **Xmin** must be set to 0°; **Xmax** must be set to 90° and **Xscl** must be set so that the grigding of the x-axis will be made properly In such a case **Xscl** being 30° would be fine. If x is an acute angle and the angle mode is radians, then **Xmin** must be set to 0; **Xmax** must be set to $\pi/2$ and **Xscl** may be set to 1.

vi. When only sines and cosines are involved, **ZoomFit** option may give a clearer graph. However, since only the x-intercepts are required, the window setting parameters **Ymin= -1** and **Ymax = 1** can give a clear view of the zeros.

vii. Find the x-intercepts using the **Calc Zero** of TI-83 Plus. However when the graph seems to be tangent to the x-axis at a certain point, you may use the **Calc Min** or **Calc Max** facilities but you should make sure that the y-coordinate of the minimum or maximum point is zero.

viii. Any value like **-6.61E -10** or **7.2E -11** can be interpreted as 0 as they mean **-6.6x10^{-10}** and **7.2x10^{-11}** respectively.

1. How many solutions does the following equation have between 0° and 360°?

$$\sec^2 x - \frac{\sin x}{\cos x} = 1$$

Solution:

Mode: Degrees

$$\sec^2 x - \frac{\sin x}{\cos x} - 1 = 0$$

 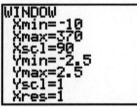 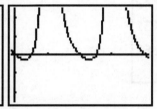

Answer: The graph intersects with the x-axis 5 times in the given interval therefore there are 5 solutions.

2. $\cos(33°) = \tan x° \Rightarrow x = ?$ (x is an acute angle)

Solution:

Mode: Degrees

$\cos(33°) - \tan x° = 0$

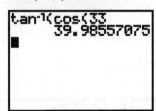

or

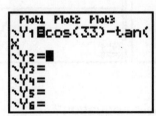  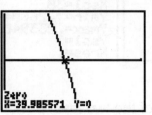

Answer: 39.99°

3. $0 < x < \frac{\pi}{4}$ and $\tan(4x) = 3$. What is x and what is tanx?

Solution:

Mode: Radians

tan(4x) – 3 = 0

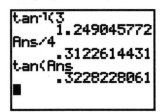

or

Answer: x= 0.312 and tanx= 0.323

4. sin(120°-n)=sin50° and n is an acute angle $\Rightarrow$ n=?

Solution:

Mode: Degrees

sin(120°-n) - sin50° = 0

   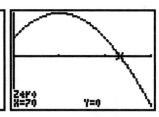

Answer: 70°

5. What is the sum of the two least positive solutions of the following equation?

sin(10x)=-cos(10x)

Solution:

Mode: Radians

Advanced Calculation and Graphing Techniques with the TI – 83 Plus Graphing Calculator

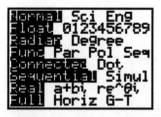

sin(10x) + cos(10x) = 0

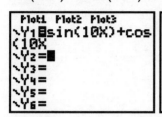 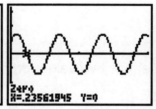 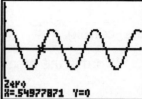

Answer: 0.24 + 0.55 = 0.79

6. cos(2x) = 2sin(90°-x). What are all possible values of x between 0° & 360°?

   Solution:

   Mode: Degrees

   cos(2x) - 2sin(90°-x) = 0

   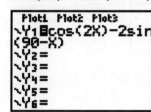 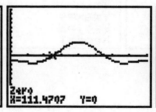 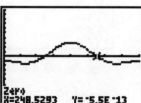

   Answer: 111.47°, 248.53°

7. $\frac{1}{4}\sin^2(2x) + \sin^2(x) + \cos^4(x) = 1$

   If x is positive and less than 2π, how many different values can x have?

   Solution:

   Mode: Radians

   $\frac{1}{4}\sin^2(2x) + \sin^2(x) + \cos^4(x) - 1 = 0$

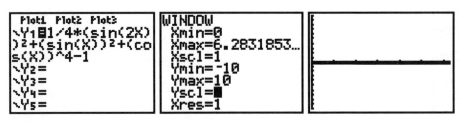

Answer: x has infinitely many values between $(0, 2\pi)$

8. $\dfrac{1}{\cot(5x)} = -2$

   What is the smallest positive value for x?

   Solution:

   Mode: Radians

   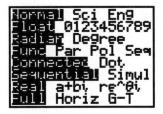

   $\dfrac{1}{\cot(5x)} + 2 = 0$

   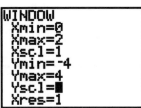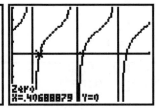

   Answer: 0.41

9. $\dfrac{8\sin(2\theta)}{1-\cos(2\theta)} = \dfrac{4}{3}$ and θ is between $0°$ and $180°$. What is θ?

   Solution:

   Mode: Degrees

   $\dfrac{8\sin(2\theta)}{1-\cos(2\theta)} - \dfrac{4}{3} = 0$

 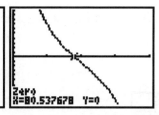

Answer: 80.53

10. $\dfrac{\sin x + \cos 36°}{\cos\dfrac{4\pi}{3} - \sin(-90°)} = 0$ and x is between 90° and 270° ⇒ x=?

Solution:

Mode: Degrees

  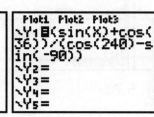 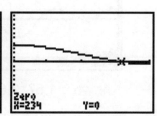

Answer: 234°

11. $\dfrac{\sin\theta}{\cos\theta - 1} = -\sqrt{3}$ ⇒ If θ is an acute angle, θ=?

Solution:

Mode: Radians

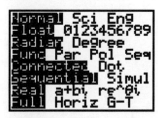

$\dfrac{\sin\theta}{\cos\theta - 1} + \sqrt{3} = 0$

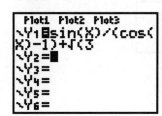 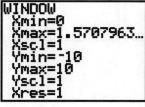 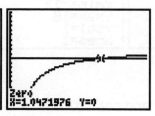

Answer: 1.05

12. $2\sin x + \cos(2x) = 2\sin^2 x - 1$ and $0 \leq x < 2\pi \Rightarrow x = ?$

    Solution:

    Mode: Radians

    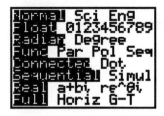

    $2\sin x + \cos(2x) - 2\sin^2 x + 1 = 0$

    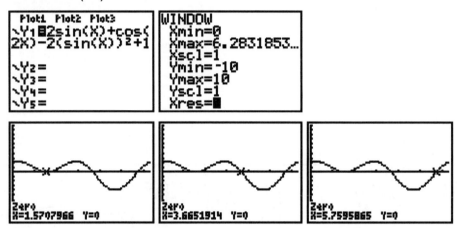

    Answer: 1.57, 3.67, 5.76

13. $\cos(130° - 2x) = \sin(70° - 3x)$ and x is an acute angle. What is x?

    Solution:

    Mode: Degrees

    $\cos(130° - 2x) - \sin(70° - 3x) = 0$

    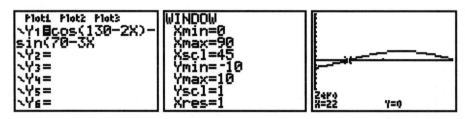

    Answer: 22°

14. x is in quadrant 3 and $\cot(120° - x) = \dfrac{1}{\tan x} \Rightarrow x = ?$

    Solution:

Mode: Degrees

$$\cot(120° - x) - \frac{1}{\tan x} = 0$$

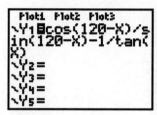 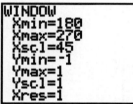 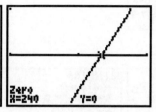

Answer: 240°

15. $\left. \begin{array}{l} \dfrac{\sin(2\theta)}{2} = \dfrac{1}{4} \\ 0° \leq \theta < 360° \end{array} \right\}$ What is θ?

Solution:

Mode: Degrees

$$\frac{\sin(2\theta)}{2} - \frac{1}{4} = 0$$

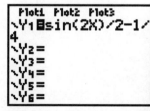 

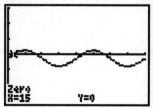 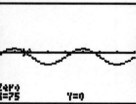 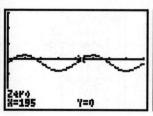 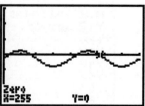

Answer: 15°, 75°, 195°, 255°

16. $\sec\theta \cdot \csc\theta = 4$ ⎫
    $0° \leq \theta < 360°$ ⎬ $\theta = ?$

Solution:

Mode: Degrees

$\sec\theta \cdot \csc\theta - 4 = 0$

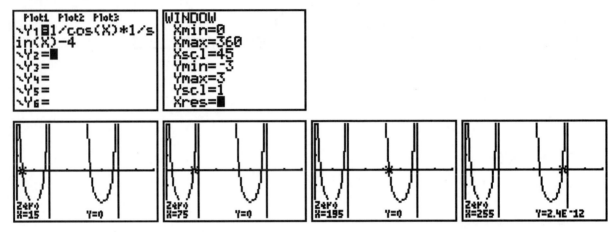

Answer: 15°, 75°, 195°, 255°

17. $0° \leq x < 90°$ ⎫
    $\tan(4x) = 1$ ⎬ $x = ?$

Solution:

Mode: Degrees

$\tan(4x) - 1 = 0$

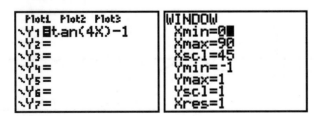

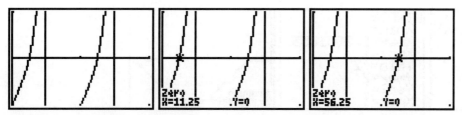

Answer: 11.25° and 56.25°

18. $\tan(6x) = \sqrt{3}$ and x is an acute angle $\Rightarrow$ x=?

   Solution:

   Mode: Radians

   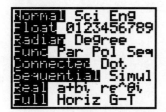

   $\tan(6x) - \sqrt{3} = 0$

   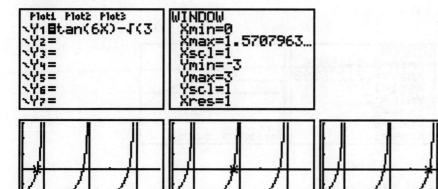

   Answer: 0.17, 0.70, 1.22

19. $\left.\begin{array}{c}\dfrac{\sqrt{3}}{2}\cos x + \dfrac{1}{2}\sin x = 1 \\ 0 \leq x < 2\pi\end{array}\right\} \Rightarrow$ x=?

   Solution:

   Mode: Radians

   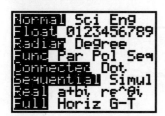

   $\dfrac{\sqrt{3}}{2}\cos x + \dfrac{1}{2}\sin x - 1 = 0$

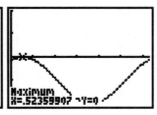

Answer: 0.52

20. $2\sin^2 x = 3(1+\cos x) - \dfrac{1}{2}$ and x is in 3rd quadrant. What is x in radians?

Solution:

Mode: Radians

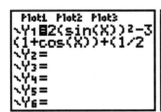

$2\sin^2 x - 3(1+\cos x) + \dfrac{1}{2} = 0$

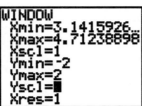 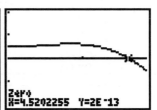

Answer: 4.52

21. cosx cos45° - sinx sin45° = -1 and x is an obtuse angle ⇒ x=?

Solution:

Mode: Degrees

cosx cos45° - sinx sin45° + 1 = 0

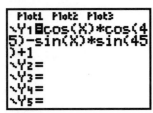 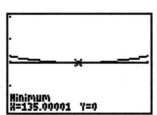

Answer: 135°

Advanced Calculation and Graphing Techniques with the TI – 83 Plus Graphing Calculator

22. $\left.\begin{array}{c} \sin x \sec x = \sqrt{3} \\ 0 \leq x < 2\pi \end{array}\right\} \Rightarrow x = ?$

Solution:

Mode: Radians

$\sin x \sec x - \sqrt{3} = 0$

 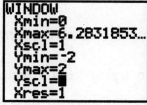 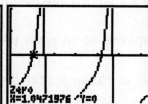 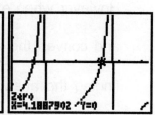

Answer: 1.05, 4.19

Advanced Calculation and Graphing Techniques with the TI – 83 Plus Graphing Calculator

5.7 Inverse Trigonometric Equations

When solving an inverse trigonometric equation in the form **f(x)=g(x)**, perform the following steps:

i. Write the equation in the form: **f(x)-g(x)=0**.

ii. While writing the trigonometric expressions please observe the rules given in parts 1.17 and 1.18 that involve the trigonometric functions that are built in TI or otherwise.

iii. Plot the graph of **y=f(x)-g(x)**.

iv. Set the **angle mode to radians or degrees** depending on which angle measure is used in the question. If no degree signs (like in 90°) are used then the mode should be radians. However when exact values are required you may wish to solve the equation in degrees and convert the answer to radians using the following formula $\frac{R}{\pi} = \frac{D}{180°}$. In such a case finding the answer in radians and then trying to find which answer choice matches this answer can also be an option; while doing so you may directly replace π with 180°.

v. Find the x-intercepts using the **Calc Zero** of TI-83 Plus. However when the graph seems to be tangent to the x-axis at a certain point, you may use the **Calc Min** or **Calc Max** facilities but you should make sure that the y-coordinate of the minimum or maximum point is zero.

vi. Any value like **-6.61E -10** or **7.2E -11** can be interpreted as 0 as they mean **-6.6x10 $^{-10}$** and **7.2x10 $^{-11}$** respectively.

1. Solve for x: $\cos^{-1}(2x - 2x^2) = \frac{2\pi}{3}$

    Solution:

    Mode: Radians

    $$\cos^{-1}(2x - 2x^2) - \frac{2\pi}{3} = 0$$

     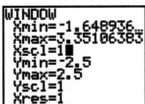 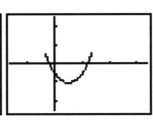

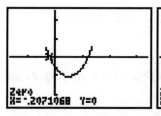 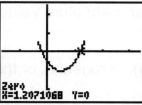

Answer: -0.207 or 1.207

2. $\sin^{-1}(x) = 3\operatorname{Arccos} x \Rightarrow x = ?$

Solution:

Mode: Radians

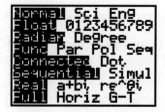

$\sin^{-1}(x) - 3\operatorname{Arccos} x = 0$

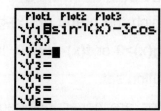 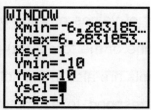 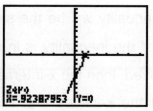

 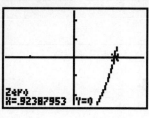

Answer: 0.92

3. Find B in degrees using the following system of equations.

$$\left.\begin{array}{l} A = \operatorname{Arctan}\left(\dfrac{-5}{12}\right) \\ A + B = 300° \end{array}\right\}$$

Solution:

Mode: Degrees

 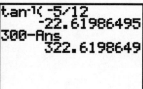

Answer: **322.62°**

Advanced Calculation and Graphing Techniques with the TI – 83 Plus Graphing Calculator

5.8 Polynomial, Algebraic and Absolute Value Inequalities

When solving an inequality in the form **f(x)<g(x)**, or **f(x)≤g(x)**, or **f(x)>g(x)**, or **f(x)≥g(x)** perform the following steps:

i. Write the inequality in the form: **f(x)-g(x)<0** or **f(x)-g(x)≤0** or **f(x)-g(x) >0** or **f(x)-g(x)≥0**.

ii. Plot the graph of **y=f(x)-g(x)**.

iii. Find the x-intercepts using the **Calc Zero** of TI-83 Plus. However when the graph seems to be tangent to the x-axis at a certain point, you may use the **Calc Min** or **Calc Max** facilities but you should make sure that the y-coordinate of the minimum or maximum point is zero.

iv. Any value like **-6.61E -10** or **7.2E -11** can be interpreted as 0 as they mean -6.6×10^{-10} and 7.2×10^{-11} respectively.

v. The solution of the inequality will be the set of values of x for which the graph of f(x)-g(x) lies below the x axis if the inequality is in one of the forms **f(x)-g(x)<0** or **f(x)-g(x)≤0**. The solution of the inequality will be the set of values of x for which the graph of f(x)-g(x) lies above the x axis if the inequality is in one of the forms **f(x)-g(x)>0** or **f(x)-g(x)≥0**. If ≤ or ≥ symbols are involved, then the x-intercepts are also in the solution set.

vi. Please note that the x-values that correspond to asymptotes are never included in the solution set.

Find the solution sets of the following:

1. $x^2-8x+7<0$

    Solution:

    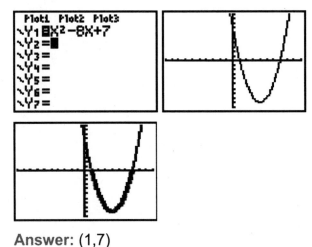

    Answer: (1,7)

2. $\dfrac{x}{x-3} > 4$

   Solution:

   $\dfrac{x}{x-3} - 4 > 0$

   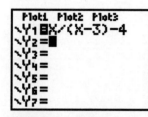 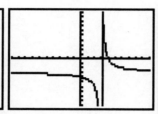 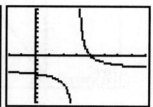

   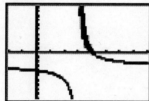

   Answer: (3,4)

3. $\dfrac{|x-2|}{x} > 3$

   Solution:

   $\dfrac{|x-2|}{x} - 3 > 0$

   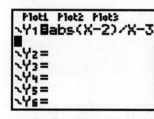 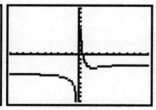  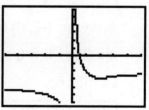

   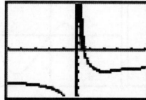

   Answer: (0, 0.5)

4. $f(x) = x + \sqrt{2x+1}$ and $f(x) \le 4$.

   Solution:

   $x + \sqrt{2x+1} - 4 \le 0$

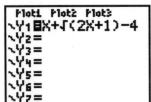 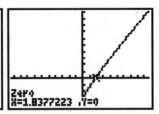

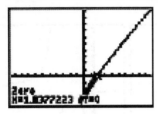

Answer: [-0.5, 1.84]

5. x(x-1)(x+2)(x-3) < 0

    Solution:

    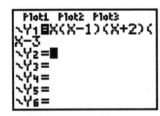 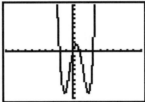

    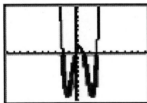

    Answer: (-2, 0) or (1,3)

6. x(x-2)(x+1) > 0

    Solution:

     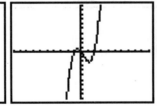  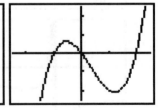

    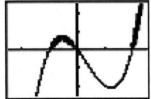

    Answer: (-1, 0) or (2,∞)

7. $x^2(x-2)(x+1) \geq 0$

    Solution:

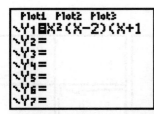  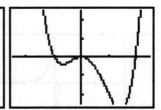

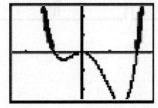

Answer: (-∞, -1] or {0} or [2, ∞)

8. $\dfrac{x+2}{x} < 4$

    Solution:

    $\dfrac{x+2}{x} - 4 < 0$

    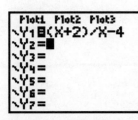 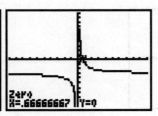

    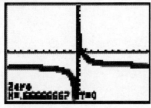

    Answer: (-∞, 0) or (0.67, ∞)

9. $4x^2 - x < 3$

    Solution:

    $4x^2 - x - 3 < 0$

    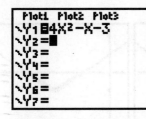 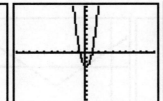 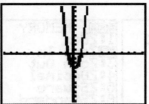

    Answer: (-0.75, 1)

10. $\dfrac{(x+1)^2}{x^3} > 0$

Solution:

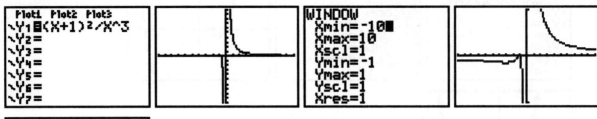

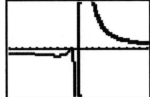

Answer: $(0, \infty)$

11. $|2x + 5| \geq 3$

    Solution:

    $|2x + 5| - 3 \geq 0$

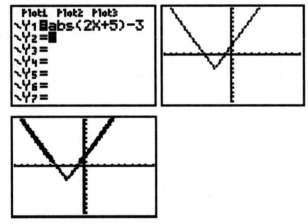

Answer: $(-\infty, -4]$ or $[-1, \infty)$

12. $|x - 2| \leq 1$

    Solution:

    $|x - 2| - 1 \leq 0$

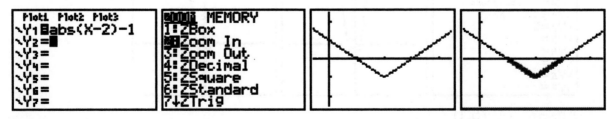

Answer: $[1, 3]$

13. $x^2 + 12 < 7x$

Solution:

$x^2 - 7x + 12 < 0$

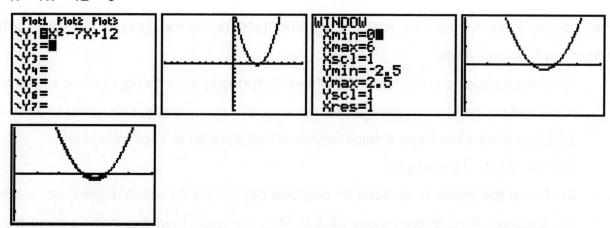

Answer: (3,4)

14. In which quadrants are the points that satisfy the following system of inequalities?

    $y < -(x-2)^2 - 1$

    $y \geq 2x - 7$

    Solution:

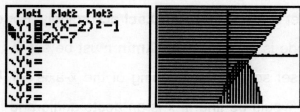

Answer: 3rd and 4th quadrants.

Advanced Calculation and Graphing Techniques with the TI – 83 Plus Graphing Calculator

5.9 Trigonometric Inequalities

When solving a trigonometric inequality in the form **f(x)<g(x)**, or **f(x)≤g(x)**, or **f(x)>g(x)**, or **f(x)≥g(x)** perform the following steps:

i. Write the inequality in the form: **f(x)-g(x)<0** or **f(x)-g(x)≤0** or **f(x)-g(x) >0** or **f(x)-g(x)≥0**.

ii. While writing the trigonometric expressions please observe the rules given in parts 1.17 and 1.18 that involve the trigonometric functions that are built in TI or otherwise.

iii. Plot the graph of **y=f(x)-g(x)**.

iv. Set the **angle mode to radians or degrees** depending on which angle measure is used in the question. If no degree signs (like in 90°) are used then the mode should be radians. However when exact values are required you may wish to solve the equation in degrees and convert the answer to radians using the following formula $\frac{R}{\pi} = \frac{D}{180°}$. In such a case finding the answer in radians and then trying to find which answer choice matches this answer can also be an option; while doing so you may directly replace π with 180°

v. If x is limited to a certain interval then set **Xmin**; **Xmax** and **Xscl** accordingly. For example, if x is an acute angle and the angle mode is degrees, then **Xmin** must be set to 0°; **Xmax** must be set to 90° and **Xscl** must be set so that the grigding of the x-axis will be made properly In such a case **Xscl** being 30° would be fine. If x is an acute angle and the angle mode is radians, then **Xmin** must be set to 0; **Xmax** must be set to $\pi/2$ and **Xscl** may be set to 1.

vi. When only sines and cosines are involved, **ZoomFit** option may give a clearer graph. However, since only the x-intercepts are required, the window setting parameters **Ymin= -1** and **Ymax = 1** can give a clear view of the zeros.

vii. Find the x-intercepts using the **Calc Zero** of TI-83 Plus. However when the graph seems to be tangent to the x-axis at a certain point, you may use the **Calc Min** or **Calc Max** facilities but you should make sure that the y-coordinate of the minimum or maximum point is zero.

viii. Any value like **-6.61E -10** or **7.2E -11** can be interpreted as 0 as they mean -6.6×10^{-10} and 7.2×10^{-11} respectively.

ix. The solution of the inequality will be the set of values of x for which the graph of f(x)-g(x) lies below the x axis if the inequality is in one of the forms **f(x)-g(x)<0** or **f(x)-g(x)≤0**. The solution of the inequality will be the set of values of x for which the graph of f(x)-g(x) lies above the x axis if the inequality is in one of the forms **f(x)-g(x)>0** or **f(x)-g(x)≥0**. If ≤ or ≥ symbols are involved, then the x-intercepts are also in the solution set.

x. Please note that the x-values that correspond to asymptotes are never included in the solution set.

1. x < cosx

   What is the solution set of the above inequality?

   Solution:

   x - cosx < 0

   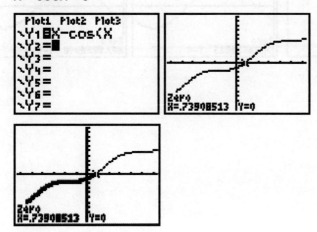

   Answer: (-∞, 0.74)

2. sin(2x) > sinx

   Find the set of values of x that satisfy the above inequality in the interval 0<x<2π.

   Solution:

   sin(2x) - sinx > 0

   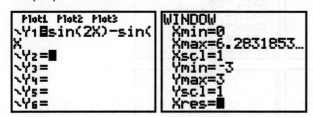

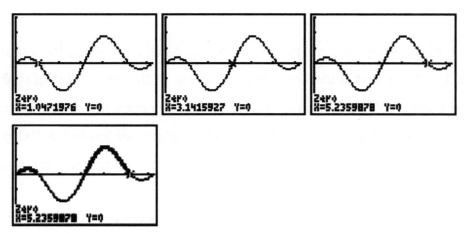

Answer: (0, 1.05) or (3.14, 5.24)

3. If x is between 0 and 2π, what will be the set of x values for which sinx < cosx?

   Solution:

   sinx - cosx < 0

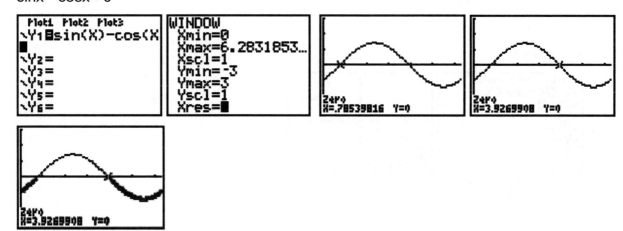

   Answer: (0, 0.79) or (3.93, 2π)

4. cos(2x) ≥ cosx

   Find the set of values of x that satisfy the above inequality in the interval $0 \leq x \leq 360°$.

   Solution:

   cos(2x) - cosx ≥ 0

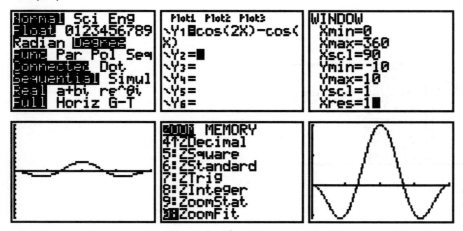

Advanced Calculation and Graphing Techniques with the TI – 83 Plus Graphing Calculator

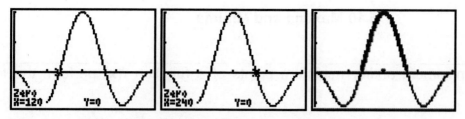

Answer: [120°, 240°] or {0°,360°}

Advanced Calculation and Graphing Techniques with the TI – 83 Plus Graphing Calculator

5.10 Maxima and Minima

When solving for the maximum and/or minimum points of a function **f(x)** perform the following steps:

i. If x is limited to a certain interval then set **Xmin**; **Xmax** and **Xscl** accordingly, otherwise use Zstandard facility while graphing **y=f(x)**.

ii. Use the **Calc Min** or **Calc Max** facilities to find the minimum and maximum point(s). However if the minimum or maximum points are at one or both of the ends of the interval, then find these points by using the **Calc Value** facility; while doing so, use the x-coordinates of the endpoints of the interval.

iii. Any value like **-6.61E -10** or **7.2E -11** can be interpreted as 0 as they mean **-6.6x10^{-10}** and **7.2x10^{-11}** respectively.

1. $f(x)=2x^2+1$ is defined in the interval $-3 \leq x \leq 3$. find minimum value of f(x).

    Solution:

    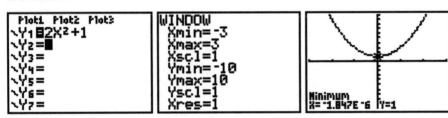

    Answer: 1

2. $f(x)=|3x+1|-1$ Find minimum value of f(x).

    Solution:

    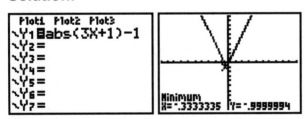

    Answer: -1

3. $f(x)=-|x|+3$ and $-2 \leq x \leq 4$ Find minimum value of f(x) and the x value where this minimum occurs.

Solution:

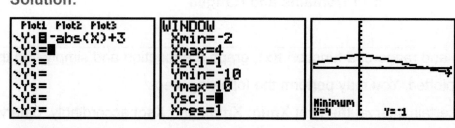

Answer: (4, -1)

4. $y = \sqrt[3]{9 - x^2}$

Find maximum value of y.

Solution:

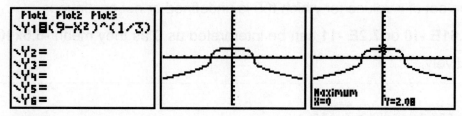

Answer: 2.08

5.11 Domains and Ranges

When finding the domain and range of a function **f(x)**, graph the function and simply find the set of x values for which f(x) is plotted. You may perform the following steps:

i. If x is limited to a certain interval then set **Xmin**; **Xmax** and **Xscl** accordingly, otherwise use Zstandard facility while graphing **y=f(x)**.

ii. Use the **Calc Zero, Calc Value**, **Calc Min,** or **Calc Max** facilities to find the zeros, minima and maxima.

iii. When asymptotes or discontinuities are involved, you may use the **TBLSET** and **TABLE** facilities to find the set of x values for which f(x) is undefined or not continuous.

iv. Any value like **-6.61E -10** or **7.2E -11** can be interpreted as 0 as they mean **-6.6x10 $^{-10}$** and **7.2x10 $^{-11}$** respectively.

1. Find the domain of $f(x)= \log \sqrt{2x^2 - 15}$.

   Solution:

   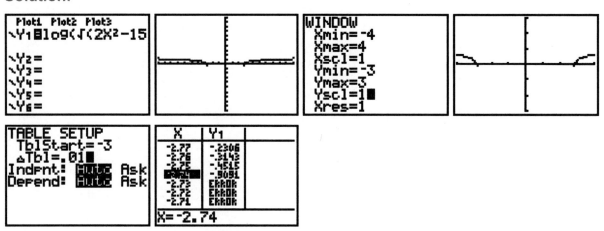

   Answer: Because of the even symmetry that the graph has and using the table, we deduce that the domain of the function is as follows: x<-2.73 or x>2.73. This can also be stated as |x|>2.73. Please note that the answer is accurate to the nearest hundredth.

2. Find domain and range of the function $y = x^{-4/3}$

   Solution:

   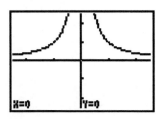

   Answer: Domain: x≠0; Range: y>0.

3. Find the domain and range of the function $f(x) = 4 - \sqrt{2x^3 - 16}$

   Solution:

   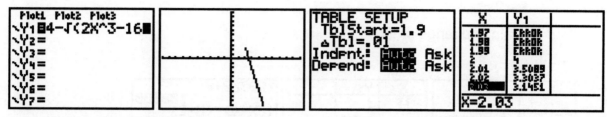

   Answer: Domain: $x \geq 2$; Range: $y \leq 4$.

4. $f(x) = 2x^2 + 5x + 2$

   $g(x) = 4x^2 - 4$

   In order that $\left(\dfrac{f}{g}\right)(x)$ be a function what must be excluded from the domain?

   Solution:

   $g(x)$ is the denominator of the given polynomial which must be other than zero.

   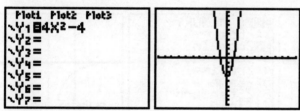

   Answer: The values that must be excluded from the domain are: -1 and 1.

5. Find range of $y = 8 - 2x - x^2$

   Solution:

   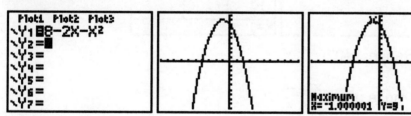

   Answer: $y \leq 9$

6. $f(x) = \log(\sin x)$

   Find domain of $f(x)$.

   Solution:

   In order that $f(x)$ be a function $\sin(x)$ must be positive.

   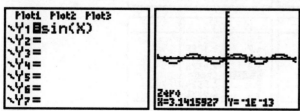

Answer: $2k\pi < x < (2k+1)\pi$ where k is any integer.

7. $f(x) = \dfrac{3x+4}{x+2}$

   Find domain and range of f(x)

   Solution:

   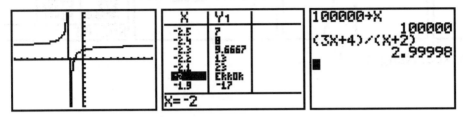

   Answer: Domain: $x \neq -2$; Range: $y \neq 3$.

8. $f(x) = \dfrac{x+1}{2x-2}$

   What value(s) must be excluded from the domain of f(x) and what is the range of f(x)?

   Solution:

   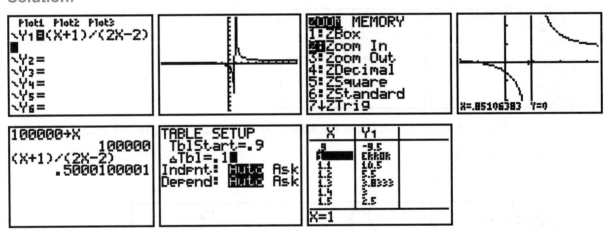

   Answer: Domain: $x \neq 1$; Range: $y \neq 0.5$.

9. What is the domain and range of $y = \dfrac{x^2-4}{x^2-2x}$?

   Solution:

   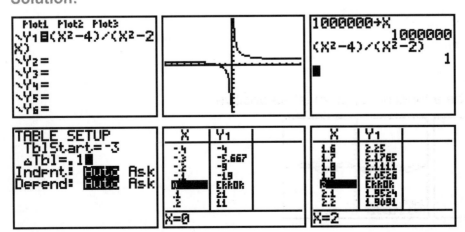

Answer: Domain: All real numbers except 0 and 2; Range: $y \neq 1$.

10. Domain of f(x) is given by $x^2+3x-4<0$ and $f(x)=x^2+4x+5$.

    Find range of f(x).

    Solution:

    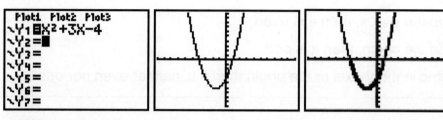

    Domain: $-4 < x < 1$

    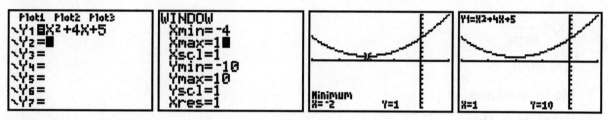

    Answer: $1 \leq y < 10$

11. Find domain and range of $y = \sqrt{x^2 - 9}$.

    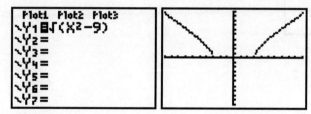

    Answer: Domain: $x \leq -3$ or $x \geq 3$; range: $y \geq 0$.

12. Find domain and range of $y = \sqrt{9 - x^2}$.

    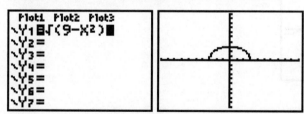

    Answer: Domain: $-3 \leq x \leq 3$; Range: $0 \leq y \leq 3$.

5.12 Evenness And Oddness

When finding whether a function **f(x)** is **even, odd** or **neither**, graph the function and simply check the symmetry.

i.   If f(x) is symmetric in the y-axis, then it is even.
ii. If f(x) is symmetric in the origin, then it is odd.
iii. If f(x) is not symmetric in the y-axis or the origin then it is neither even nor odd.

State whether each of the following functions are even, odd, or neither.

1. $f(x) = \dfrac{1}{\sec x}$

    Solution:

    Mode: Radians

    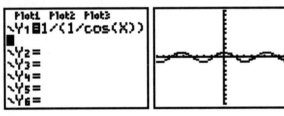

    Answer: Even (symmetric in the y-axis)

2. f(x) = cosx

    Solution:

    Mode: Radians

    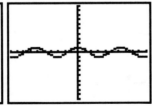

    Answer: Even

3. $f(x) = \dfrac{1}{\csc(x)}$

    Solution:

    Mode: Radians

    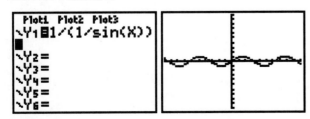

Answer: Odd (symmetric in the origin)

4. $f(x) = \sin x$

   Solution:

   Mode: Radians

   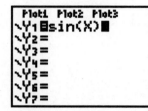 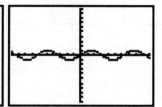

   Answer: Odd

5. $f(x) = \sin x + 1$

   Solution:

   Mode: Radians

   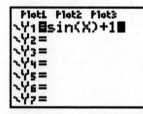 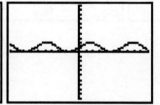

   Answer: Neither odd nor even (no symmetry in the y-axis or the origin)

6. $f(x) = \dfrac{1}{x}$

   Solution:

   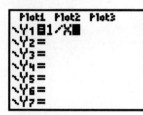 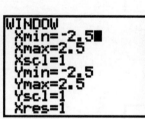 

   Answer: Odd

7. $f(x) = |x|$

   Solution:

    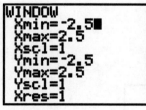 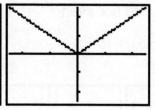

   Answer: Even

8. $f(x) = \log(x^2)$

Solution:

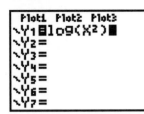 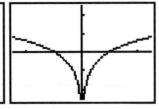

Answer: Even

9. f(x) = -x² + sinx

   Solution:

   Mode: Radians

    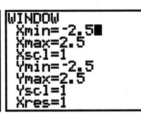 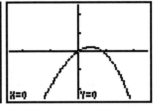

   Answer: Neither odd nor even

10. f(x) = x⁴ - 3x² + 5

    Solution:

    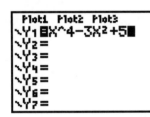 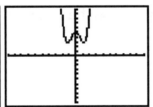

    Answer: Even

11. f(x) = 3x³ + 5

    Solution:

     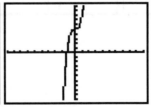

    Answer: Neither odd nor even

12. f(x) = 12x⁶ + 4x⁴ - 13x²

    Solution:

    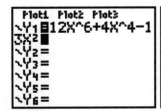 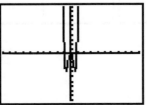

Answer: Even

13. $f(x) = -x^5 - 8x^3 + 12x$

Solution:

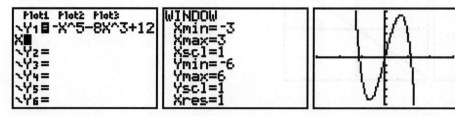

Answer: Odd

14. $f(x) = x^3$

Solution:

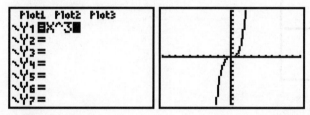

Answer: Odd

15. $f(x) = 3x^4 + 2x^2 - 8$

Solution:

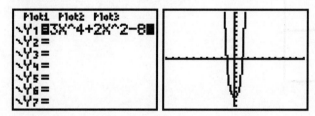

Answer: Even

16. $y = 2$

Solution:

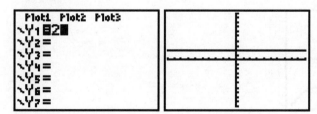

Answer: Even

Advanced Calculation and Graphing Techniques with the TI – 83 Plus Graphing Calculator

Unauthorized copying or reuse of any part of this page is illegal.

17. y = x

    Solution:

    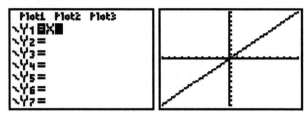

    Answer: Odd

18. f(x) = x³-1

    Solution:

    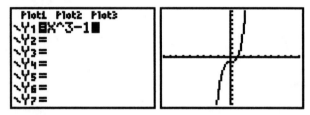

    Answer: Neither odd nor even

19. f(x) = x²-1

    Solution:

    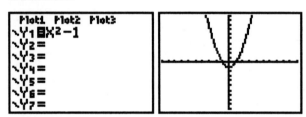

    Answer: Even

20. f(x) = -x + sinx

    Solution:

    Mode: Radians

    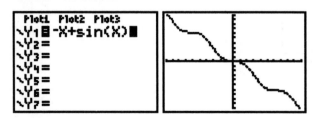

    Answer: Odd

21. f(x) = -x

    Solution:

Advanced Calculation and Graphing Techniques with the TI – 83 Plus Graphing Calculator

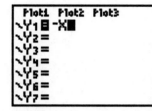 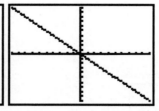

Answer: Odd

22. $f(x) = x^2$

Solution:

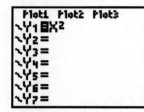 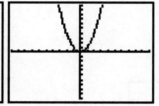

Answer: Even

23. $f(x) = \dfrac{1}{x^2}$

Solution:

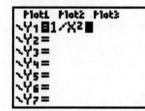 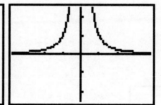

Answer: Even

24. $f(x) = 2x^4$

Solution:

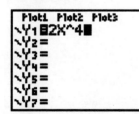 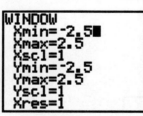 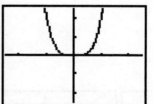

Answer: Even

25. $f(x) = x^3 + 1$

Solution:

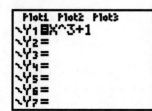 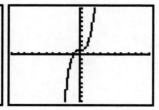

Answer: Neither odd nor even

26. $f(x) = \dfrac{x}{x-2}$

Solution:

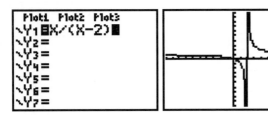

Answer: Neither odd nor even

27. $f(x) \; x^3 + x$

Solution:

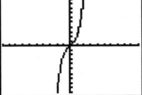

Answer: Odd

28. $f(x) = \sin(x)$

Solution:

Mode: Radians

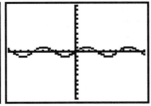

Answer: Odd

29. $f(x) = \sqrt{x^2} + 1$

Solution:

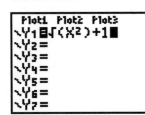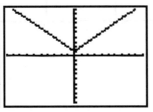

Answer: Even

30. $f(x) = \cos x$

$g(x) = 2x+1$

Solution:

Mode: Radians

i) f(x) . g(x)

 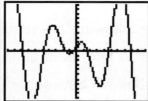

Answer: Neither odd nor even

ii) f(g(x))

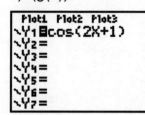  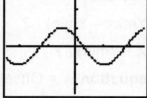

Answer: Neither odd nor even

iii) g(f(x))

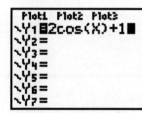 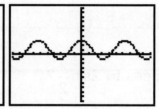

Answer: Even

5.13 Graphs of Trigonometric Functions

i. Most of the time one or more of the following are required concerning the graphs of the trigonometric functions. In order to find them all it is usually enough to find two adjacent maxima and the minimum point in between.

Period = The x-distance between two identical points in a periodic function; for example two adjacent maxima, minima or zeros.

Frequency = 1 / Period

Amplitude = (Ymax − Ymin) / 2

Offset = (Ymax + Ymin) / 2

Axis of wave equation: y = Offset

ii. **y-intercept** is the point whose x-coordinate is zero.

iii. Use the window, **Calc Min**, **Calc Max**, **Calc Value**, and **Calc Zero** facilities in order to perform the above calculations.

1. As x increases from 0 to π, what happens to $2\sin\dfrac{x}{2}$?

   Solution:

   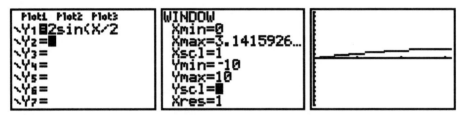

   Answer: The function increases throughout in the given interval.

2. What is the amplitude, Axis of wave and offset of y=5sin(x)+12cos(x)−2?

   Solution:

   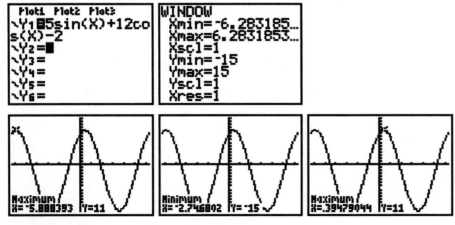

   Answer: Amplitude = (11 + 15) / 2 = 13

Advanced Calculation and Graphing Techniques with the TI – 83 Plus Graphing Calculator

Offset: (11 – 15) / 2 = -2

Axis of wave: y = -2

3. What is the maximum value of $y = \sqrt{4 + \cos^2 x}$ in the interval $\left[\dfrac{-\pi}{2}, \dfrac{\pi}{2}\right]$

   Solution:

   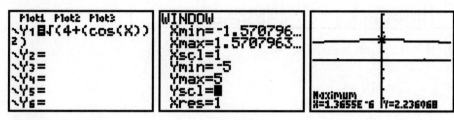

   Answer: 2.24

4. Find y intercept of the function $y = \left|\sqrt{3}\sec\left[3\left(x + \dfrac{\pi}{4}\right)\right]\right|$

   Solution:

   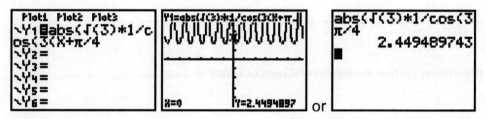

   Answer: 2.45

5. Find amplitude of the function $f(x) = -\dfrac{1}{2}\sin(x)\cos(x) + 1$

   Solution:

   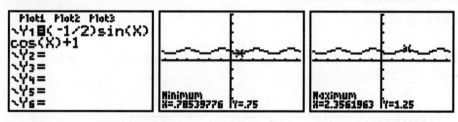

   Answer: (1.25 – 0.75) / 2 = 0.25

6. Find the primary period of $f(x) = \dfrac{\cos(2x)}{1 + \sin(2x)}$

   Solution:

   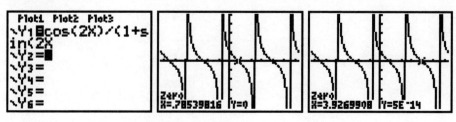

   Answer: 3.93 – 0.79 = 3.14 → π

7. Find primary period of f(x)=3sin²(2x)

Solution:

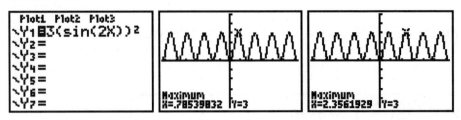

Answer: 2.36 − 0.79 = 1.57 → π / 2

8. Find y intercept of $y = \sqrt{3}\sin(x + \frac{\pi}{3})$.

Solution:

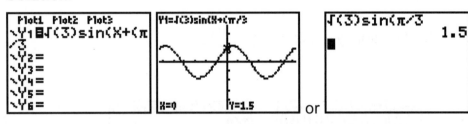

Answer: 1.5

9. What is the amplitude of the function y=3sinx+4cosx+1

Solution:

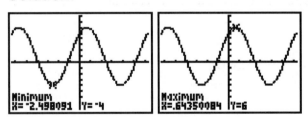

Answer: (6 − (− 4)) / 2 = 5

10. Find maximum value of the function $f(x) = \sin(\frac{x}{4})$ over the interval $0 \le x \le \frac{\pi}{3}$

Solution:

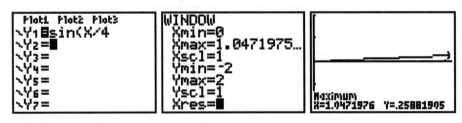

Answer: 0.26

11. Find maximum value of 4 sinx cosx

    Solution:

    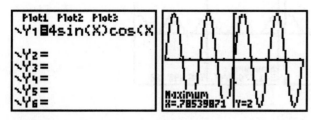

    Answer: 2

12. What happens to sinx as x increases from $-\dfrac{\pi}{4}$ to $\dfrac{3\pi}{4}$?

    Solution:

    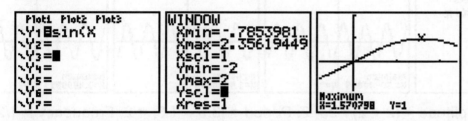

    Answer: The function increases between $\pi/4$ and $\pi/2$ and then decreases between $\pi/2$ and $3\pi/4$.

13. What is the smallest positive x intercept of $y = 2\sin\left[3(x+\dfrac{3\pi}{4})\right]$?

    Solution:

    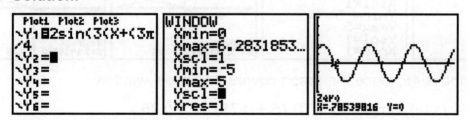

    Answer: 0.79 → $\pi/4$

14. What is the smallest positive angle that will make $y = 3 + \sin\left[3(x+\dfrac{\pi}{3})\right]$ a minimum?

    Solution:

    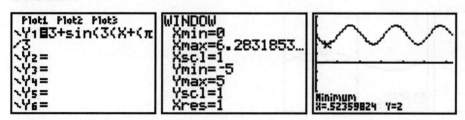

    Answer: 0.52 → $\pi/6$

15. Find amplitude of the graph of the function $y=\cos^4 x - \sin^4 x + 1$

Solution:

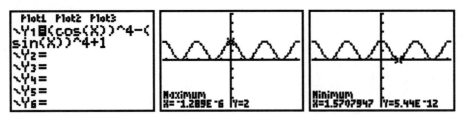

Answer: (2 − 0) / 2 = 1

16. Find amplitude, period and frequency of the following:

    1. y=2sin(πx+π)

       Solution:

       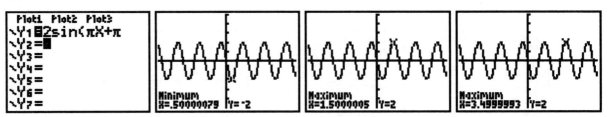

       Answer: Amplitude = (2 + 2) / 2 = 2; Period = |3.5 − 1.5| = 2; Frequency = 1/2

    2. $y = \frac{3}{4}\cos(\frac{x}{2} - \frac{\pi}{2})$

       Solution:

       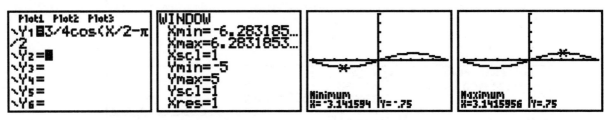

       Answer: Period is 4π, since the graph covers the whole window.

       Frequency = 1 / (4π); Amplitude = (0.75 + 0.75) / 2 = 0.75.

17. Find the coordinates of the first maximum point in the graph of $y = \sin(\frac{x}{2})$ that has a positive x-coordinate.

    Solution:

    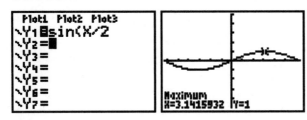

    Answer: 3.14 → π

Advanced Calculation and Graphing Techniques with the TI – 83 Plus Graphing Calculator

5.14 Miscellaneous Graphs

1. Find the point of intersection of the graphs $y=\log x$ and $y=\ln\dfrac{x}{2}$

   Solution:

   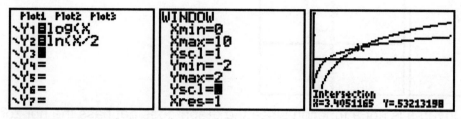

   Answer: (3.41, 0.53)

2. At how many points does the function $y=x^3+5x-2$ intersect the x axis?

   Solution:

   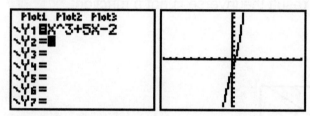

   Answer: 1 point.

3. Plot the graph of $f(x)=\dfrac{x^2-1}{x-1}$; locate the hole that the function has.

   Solution:

   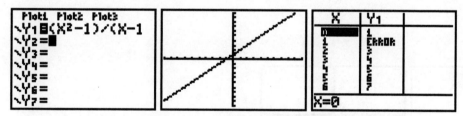

   Please note that at x=1 the function is not defined and this is not a vertical asymptote, therefore the graph has a hole at that point.

4. Find x and y intercept(s) of the graph of equation $y=(x^2-4)\ln(x^2+9)$

   Solution:

   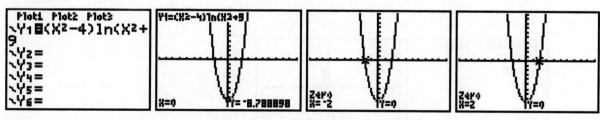

   Answer: y-intercept: (0, -8.79) , x-intercepts: (-2,0) and (2,0)

5. Determine which of the following functions has an inverse that is also a function.

   a. $y = x^2 - 3x + 5$

   Solution:

   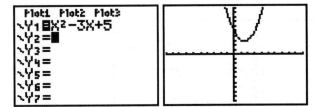

   Answer: The inverse is not a function since the graph does not pass the horizontal line test, i.e. a horizontal line that cuts a certain graph at more than one point indicating that to one y value, more than one x values correspond. Such graphs do not correspond to one to one and onto functions. As a result the inverse is not a function.

   b. $y = |x + 2| - 1$

   Solution:

   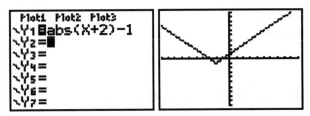

   Answer: The inverse is not a function.

   c. $y = \sqrt{16 - 9x^2}$

   Solution:

   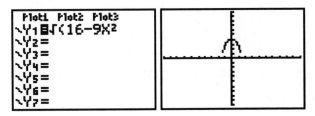

   Answer: The inverse is not a function.

   d. $y = x^3 + 5x - 2$

   Solution:

   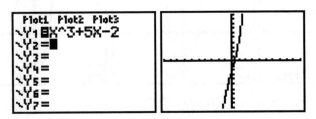

   Answer: The inverse is a function.

Advanced Calculation and Graphing Techniques with the TI – 83 Plus Graphing Calculator

Unauthorized copying or reuse of any part of this page is illegal.

6. $f(x)=2x^2+12x+3$. If the graph of $f(x-k)$ is symmetric about the y axis, what is k?

   Solution:

   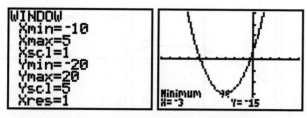

   Answer: The graph must be shifted 3 units toward right therefore k=3.

7. Find equation of the axis of symmetry of $y=3x^2-x+2$.

   Solution:

   Answer: The graph has the axis of symmetric of x = 0.17.

8. $y=2x^3+x+1$. Find the distance between the x and y intercepts of the above function.

   Solution:

   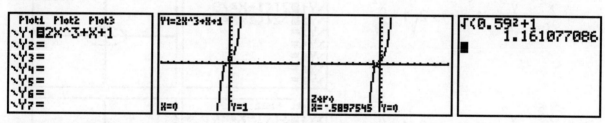

   Distance= $\sqrt{0.59^2 + 1}$ =1.161

   Answer: 1.16

9. $y=-2x^2+4x-7$. Determine the coordinates of the vertex of the parabola given above. Does the above function have a maximum or minimum? What is this value? Find the equation of the axis of symmetry also.

   Solution:

   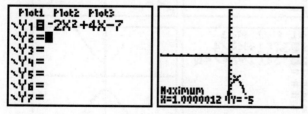

   Answer: The vertex is at (1,-5). This is also the maximum point of the graph. The graph doesn't have a minimum point since it is a downward parabola that tends to negative infinity. Equation of the axis of symmetry is x=1 and it is a vertical line.

10. $f(x) = -(x-1)^2 + 3$ and $-2 \leq x \leq 2$. Find the range of $f(x)$.

Solution:

 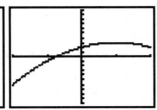 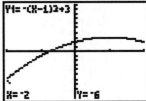

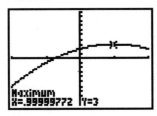

Answer: $-6 \leq y \leq 3$.

11. Plot the graphs of the following:

a. $\dfrac{x^2}{9} + \dfrac{y^2}{4} = 1$ b. $\dfrac{x^2}{4} + \dfrac{y^2}{9} = 1$ c. $\dfrac{x^2}{9} - \dfrac{y^2}{4} = 1$ d. $\dfrac{y^2}{9} - \dfrac{x^2}{4} = 1$

Solution:

a. $\dfrac{x^2}{9} + \dfrac{y^2}{4} = 1 \Rightarrow y = \pm 2\sqrt{1 - \dfrac{x^2}{9}}$

 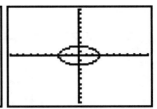

b. $\dfrac{x^2}{4} + \dfrac{y^2}{9} = 1 \Rightarrow y = \pm 3\sqrt{1 - \dfrac{x^2}{4}}$

 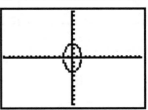

c. $\dfrac{x^2}{9} - \dfrac{y^2}{4} = 1 \Rightarrow y = \pm 2\sqrt{\dfrac{x^2}{9} - 1}$

 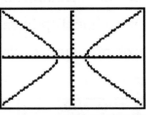

d. $\dfrac{y^2}{9} - \dfrac{x^2}{4} = 1 \Rightarrow y = \pm 3\sqrt{1 + \dfrac{x^2}{4}}$

 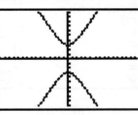

5.15 The Greatest Integer Function

The greatest integer function f(x) = [x] = [|x|] means "The greatest integer less than or equal to x". Mathematical definition for the greatest integer function is as follows:

f(x) = k if k ≤ x < k+1 and k=integer ⇒ f(x) = [x]

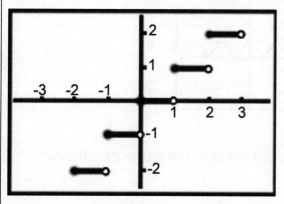

[4] = 4   [0.5]= 0   [9.76]= 9   [-3]= -3   [-8.67]= -9   [-0.32]= -1

TI Usage:

y=int(x) and style must be set to dot

1. f(x)=k where k is an integer for which k ≤ x < k+1 and g(x) = |f(x)|-f(x)+1. What is the minimum value for g(x)?

   Solution:

   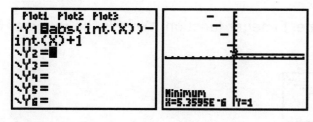

   Answer: 1

2. f(x)=|1-2x+2[x]|

   What is the period and frequency of the above function if [x] represents the greatest integer less than or equal to x? What are the maximum and the minimum values of f(x)? What is the amplitude, offset, and equation of the Axis of wave? What is the domain and range?

Solution:

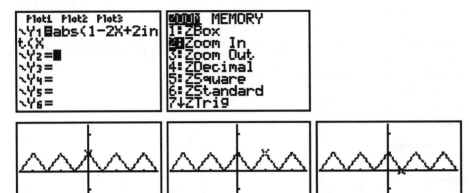

Answer: Min = 0; Max = 1

Period = | 1 – 0 | = 1 → (The distance between two adjacent maxima or minima)

Frequency = 1 → (Frequency = 1 / Period)

Amplitude = (1 – 0) / 2 = ½ → (Amplitude = (Ymax – Ymin) / 2)

Offset: (1 + 0) / 2 = ½ → (Offset = (Ymax + Ymin) / 2)

Axis of wave: y = ½ → (Axis of wave equation is y = Offset)

Domain: R, Range: $0 \le y \le 1$.

3. g(x)=[x]-2x+1 what is the period of g(x)?

Solution:

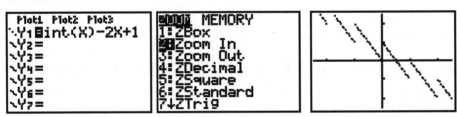

Answer: The function is not periodic.

4. f(x)=[x] where [x] represents the greatest integer function. What is the range of f(x)?

Solution:

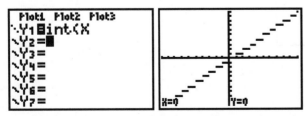

Answer: The range is all integers. The domain is all real numbers.

5. [4.6]- [-5.4]+2[0.3]+ [4]- [0]=?

Solution:

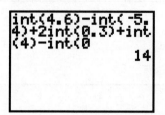

Answer: 14

5.16 Parametric Graphs

Please review section 2.9.

Plot the graph and state what each curve represents.

1. $x = 4\cos\theta + 1$

   $y = 3\sin\theta - 1$

   Solution:

   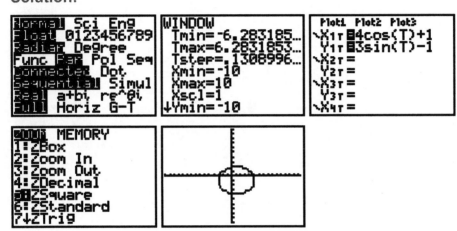

   Answer: The curve represents an ellipse.

2. $x = t^2 + t + 1$

   $y = t^2 - t + 1$

   Solution:

   Function Mode: Par(ametric), Tmin = -2π, Zoom = ZSquare

   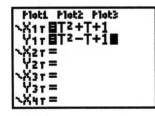 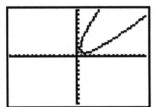

   Answer: The curve represents a parabola.

3. $x = t^3 + 2$

   $y = \dfrac{4}{3}t^3 + 1$

   Solution:

   Function Mode: Par(ametric), Tmin = -2π, Zoom = ZSquare

Advanced Calculation and Graphing Techniques with the TI – 83 Plus Graphing Calculator

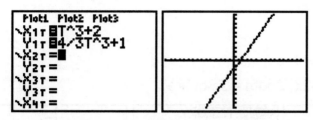

Answer: The curve represents a line.

4. $x = t^2$

   $y = 2t^2 - 1$

   Solution:

   Function Mode: Par(ametric), Tmin = -2π, Zoom = ZSquare

   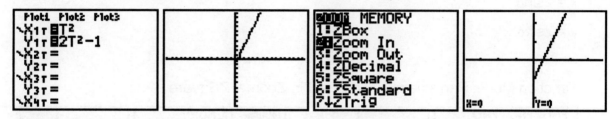

   Answer: The curve represents a portion of a line that starts at (0,-1).

5. $x = \sin\theta$

   $y = \cos\theta$

   Solution:

   Function Mode: Par(ametric), Tmin = -2π, Zoom = ZSquare

   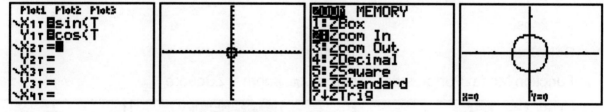

   Answer: The curve represents the unit circle.

6. $x = t$

   $y = \sqrt{4 - t^2}$

   Solution:

   Function Mode: Par(ametric), Tmin = -2π, Zoom = ZSquare

   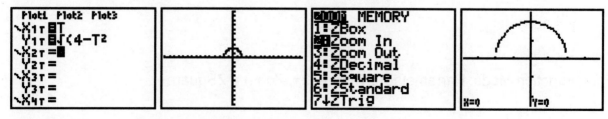

   Answer: The curve represents a semicircle.

7. $x = \sqrt{p}$

$y = \sqrt{4-p}$

Solution:

Function Mode: Par(ametric) , Tmin = -2π, Zoom = ZSquare

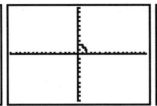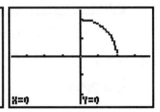

Answer: The curve represents a quarter circle.

8. $x = 2\sin\alpha$

    $y = 2\sin\alpha$

    Solution:

    Function Mode: Par(ametric) , Tmin = -2π, Zoom = ZSquare

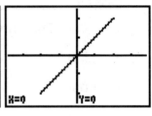

Answer: The curve represents a portion of a line ranging from (-2,-2) to (2,2).

9. $x = 2\sin\alpha$

    $y = 2\cos\alpha$

    Solution:

    Function Mode: Par(ametric) , Tmin = -2π, Zoom = ZSquare

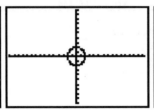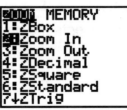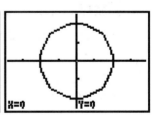

Answer: The curve represents a circle.

10. $x = 3\sin\alpha$

    $y = 4\cos\alpha$

    Solution:

    Function Mode: Par(ametric) , Tmin = -2π, Zoom = ZSquare

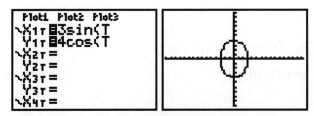

Answer: The curve represents an ellipse (a y-ellipse).

11. x = 3t+4

    y = t-6

    Solution:

    Function Mode: Par(ametric), Tmin = -2π, Zoom = ZSquare

    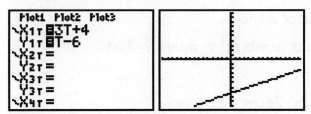

    Answer: The curve represents a line.

12. x = sin²t

    y = 3cost

    Solution:

    Function Mode: Par(ametric), Tmin = -2π, Zoom = ZSquare

    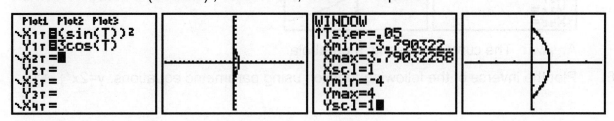

    Answer: The curve represents portion of a parabola.

13. x = t(1+t)

    y = t(-1+t)

    Solution:

    Function Mode: Par(ametric), Tmin = -2π, Zoom = ZSquare

    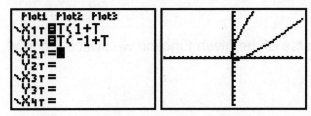

    Answer: The curve represents a parabola.

14. Using parametric equations plot the set of points (x²,y) where y=x²+1

Solution:

Function Mode: Par(ametric) , Tmin = -2π, Zoom = ZSquare

(x^2, y) where $y = x^2 + 1$ is equivalent to (x^2, x^2+1), thus:

$x = T^2$

$y = T^2 + 1$

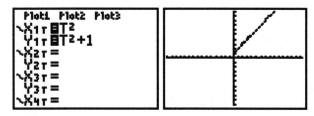

Answer: The curve represents a portion of a line.

15. Using parametric equations plot the set of points (x^2, y) where $y = 2x - 1$.

    Solution:

    Function Mode: Par(ametric) , Tmin = -2π, Zoom = ZSquare

    $x = T^2$

    $y = 2T - 1$

    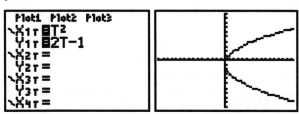

    Answer: The curve represents a parabola.

16. Plot the inverse of the following function using parametric equations. $y = 2x^3 + x + 1$

    Solution:

    Function Mode: Par(ametric) , Tmin = -2π, Zoom = ZSquare

    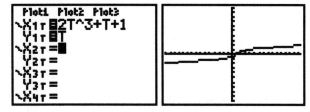

    Answer:

    Please note that in order to plot the inverse of the given function we switch x and y.

5.17 Polar Graphs

> Please review section 2.10.

1. What is the area enclosed by the following curves and the coordinate axes?

   $r = \dfrac{3}{\sin\theta}$

   $r = \dfrac{4}{\cos\theta}$

   Solution:

   Answer: 4·3=12

2. $r = \dfrac{4}{\dfrac{1}{\sec\theta} + 2\sin\theta}$ what is the area of the region bounded by the above curve and the x and y axes?

   Solution:

   Function Mode: Pol(ar)

   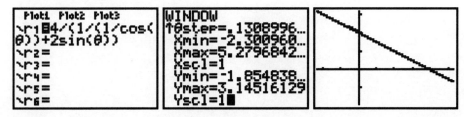

   Answer: 2·4/2=4

3. What is the area of the region that the curve r=3cosθ represents?

   Solution:

   Mode: Pol(ar)

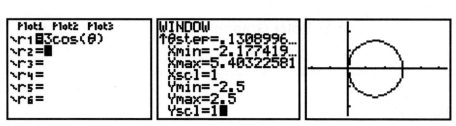

Answer: $\pi \cdot 1.5^2 = 2.25\pi = 7.07$

Advanced Calculation and Graphing Techniques with the TI – 83 Plus Graphing Calculator

5.18 Limits

i. For a function to have a limit for a given value of x=a, the right hand limit at a^+ and the left hand limit at a^- must be the same and each limit must be equal to a real number L other than infinity.

   Existence of Limit: If $\lim_{x \to a^+} f(x) = \lim_{x \to a^-} f(x) = L$ and $L \in R$ then $\lim_{x \to a} f(x) = L$

ii. Limit for a certain value of x or limit at infinity can be calculated by using the **STO**re facility of TI. What must be done is simply to store a value in x and calculate the value of the expression for this x-value.

iii. ∞ can be replaced by 100,000,000,000; and -∞ can be replaced by -100,000,000,000.

iv. Limit at a value of x other than ±∞ must be calculated as follows: If for example the limit at x=3 will be calculated, 3.000000001 (which means the right hand limit at 3^+) must be stored in x and the expression must be evaluated; then 2.999999999 (which means the left hand limit at 3^-) must be stored in x and the expression must be evaluated again. If both limits are the same, say L, then the limit is equal to L, otherwise there is no limit.

1. $\lim_{x \to \infty} \dfrac{3x^4 - 5x^3 + 8}{-4x^4 + 7x^2 + 4x + 5} = ?$

   Solution:

   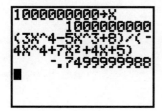

   Answer: -0.75

2. $\lim_{x \to \infty} \dfrac{6x^3 + 5x^2 - 8x}{-2x^2 + 1} = ?$

   Solution:

   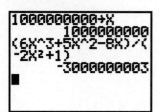

   Answer: -∞

Advanced Calculation and Graphing Techniques with the TI – 83 Plus Graphing Calculator

3. $\lim\limits_{x \to 2} \dfrac{-x^2 + 4}{x^3 + 8} = ?$

   Solution:

   ```
   2→X
                    2
   (-X²+4)/(X^3+8)
                    0
   ```

   Answer: 0

4. $\lim\limits_{x \to -\infty} \dfrac{x^3 - 27}{x^4 - 81} = ?$

   Solution:

   ```
   -1000000000→X
          -1000000000
   (X^3-27)/(X^4-81)
                -1E-9
   ```

   Answer: 0

5. $\lim\limits_{x \to 3} \dfrac{x^3 - 27}{x^4 - 81} = ?$

   Solution:

   ```
   3.0000001→X              2.999999→X
          3.0000001                 2.999999
   (X^3-27)/(X^4-81)        (X^3-27)/(X^4-81)
                  .25                .2500000417
   ```

   Answer: 0.25

6. $\lim\limits_{x \to \infty} \dfrac{6x^3 - 9x + 1}{5x^3 - 7} = ?$

   Solution:

   ```
   100000000→X
           100000000
   (6X^3-9X+1)/(5X^
   3-7)
                  1.2
   ```

   Answer: 1.2

Advanced Calculation and Graphing Techniques with the TI – 83 Plus Graphing Calculator

7. $\lim\limits_{x \to 2^+} \dfrac{3x+5}{x-2} = ?$

   Solution:

   ```
   2.0000001→X
              2.0000001
   (3X+5)/(X-2)
              110000003
   ```

   Answer: $+\infty$

8. $\lim\limits_{x \to 2} \dfrac{3x+5}{x-2} = ?$

   Solution:

   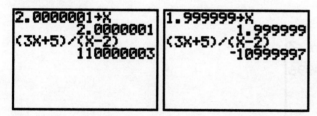

   Answer: Limit does not exist because the right hand limit and the left hand limits are not the same.

5.19 Continuity

For a function to be continuous at x=a, the right hand limit at a^+ and the left hand limit at a^- must be the same and the limit must also be equal to the value of f(x) calculated at x=a.

Continuity: If $\lim_{x \to a^+} f(x) = \lim_{x \to a^-} f(x) = f(a)$ then f(x) is continuous at x=a.

1. $f(x) = \begin{cases} \dfrac{4x^2 + 3x}{x} & x \neq 0 \\ m & x = 0 \end{cases}$

   m=? if f(x) is a continuous function.

   Solution:

   Answer: 3

2. In order to be continuous at x=2 what must $f(x) = \dfrac{x^4 - 16}{x^3 - 8}$ be defined to be equal to?

   Solution:

   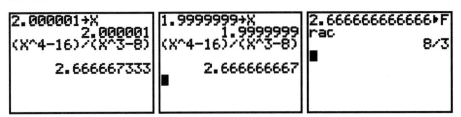

   Answer: 8/3

3. $f(x) = \begin{cases} \dfrac{6x^2 - 6}{x - 1} & x \neq 1 \\ A & x = 1 \end{cases}$

   What must be A if f(x) is a continuous function?

Solution:

```
1.0000001→X
           1.0000001
(6X²-6)/(X-1)
                  12
```

```
.99999999→X
           .99999999
(6X²-6)/(X-1)
                  12
```

Answer: 12

5.20 Horizontal and Vertical Asymptotes

$f(x) = \dfrac{P(x)}{Q(x)}$ where P(x) and Q(x) are both polynomial functions.

i. **Zero:** If $P(x_o)=0$ and $Q(x_o) \neq 0$ then f(x) has a zero at $x=x_o$.

ii. **Hole:** If $P(x_o)=0$ and $Q(x_o)=0$, and the multiplicity of x_o is the same in both polynomials, then f(x) has a hole at $x=x_o$

iii. **Vertical asymptote:** If $P(x_o) \neq 0$ but $Q(x_o)=0$, then f(x) has a vertical asymptote at $x=x_o$

iv. **Horizontal asymptote:** If the limit of $\dfrac{P(x)}{Q(x)}$ equals b as x goes to $\pm\infty$ then y=b is the horizontal asymptote.

1. Find the horizontal and vertical asymptotes as well as the domain and range of:

    a) $y = \dfrac{2x^2 - 18}{x^2 - 4}$

    Solution:

    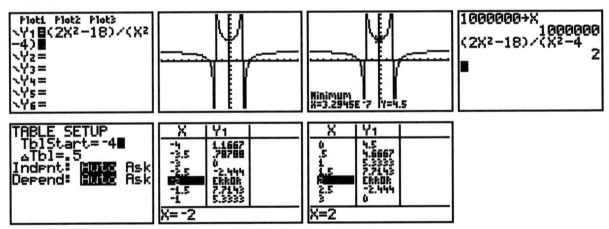

    The function apparently has two vertical asymptotes and one horizontal asymptote.

    Answer:

    Horizontal asymptote: y=2

    Vertical asymptotes: x=-2 and x=2

    Domain: All real numbers except -2 and 2

    Range: y<2 and y≥4.5.

    b) $y = \dfrac{x+2}{x^2 - 4}$

    Solution:

Advanced Calculation and Graphing Techniques with the TI – 83 Plus Graphing Calculator

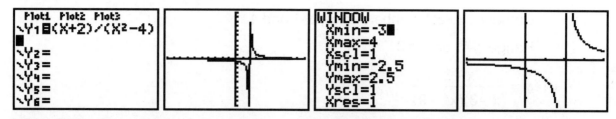

The function apparently has one vertical asymptote and one horizontal asymptote.

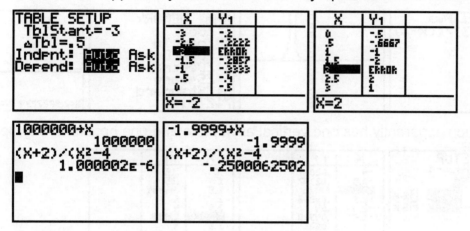

Answer:

Horizontal asymptote: y=0

Vertical asymptote: x=2

Domain: All real numbers except -2 and 2

Range: y≠0 and y≠-0.25.

c) $y = \dfrac{x^2 - 4x - 5}{x^2 - 1}$

Solution:

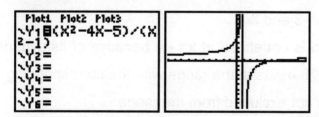

The function apparently has one vertical asymptote and one horizontal asymptote.

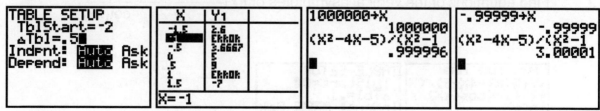

Answer:

Horizontal asymptote: y=1

Vertical asymptote: x=1

Domain: All real numbers except -1 and 1

Range: y≠1 and y≠3.

d) $y = \dfrac{x+3}{(x-3)(x^2-9)}$

Solution:

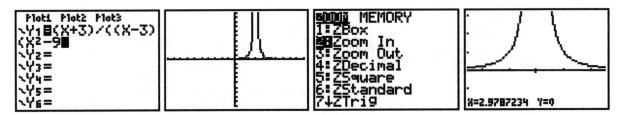

The function apparently has one vertical asymptote and one horizontal asymptote.

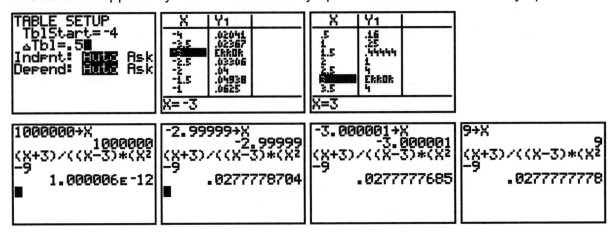

Answer:

Horizontal asymptote: y=0

Vertical asymptote: x=3

Domain: All real numbers except -3 and 3

Range: y>0 (Although the graph is undefined at x= -3, because of its symmetry about the line x=3, the y value of 0.0278 exists in the range with the corresponding x value of 9. Therefore the value 0.0278 is not excluded from the range.

2. Find equations of the vertical asymptotes of $f(x) = \dfrac{x^2+4x+3}{x+2} \cdot \dfrac{1}{\cot(\pi x)}$

Solution:

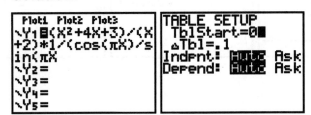

Advanced Calculation and Graphing Techniques with the TI – 83 Plus Graphing Calculator

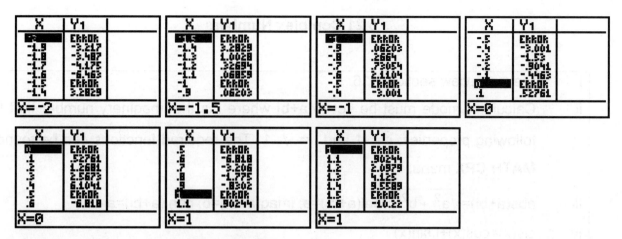

Answer: x=k/2 where k is an integer.

Advanced Calculation and Graphing Techniques with the TI – 83 Plus Graphing Calculator

5.21 Complex Numbers

> i. Please review section 1.16.
> ii. Calculator mode must be set to **a+bi** where i is the imaginary number that has the following properties: $i^2 = -1$ and $i = \sqrt{-1}$. The required functions can be found at the **MATH CPX** menu:
> iii. $abs(a+bi) = \sqrt{a^2 + b^2}$; real(a+bi)=a; imag(a+bi)=b; conj(a+bi)=a–bi.
> iv. cis(x)=cos(x)+i.sin(x)
> v. e^{ix} = cos(x)+i.sin(x)

1. $4(\text{cis}70°)^4 = ?$

    Solution:

     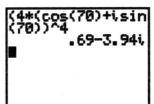

    Answer: 0.69 - 3.94i

2. $f(x) = 3x^5 - 2x^3 + 8x - 2$

    $f(i) = ?$

    Solution:

    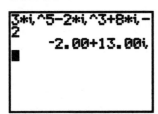

    Answer: -2+13i

3. If n is an arbitrary positive integer then $i^{4n+5} + i^{4n+6} + i^{4n+7} + i^{4n+8} = ?$

    Solution:

    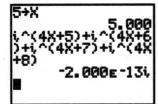 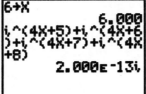

    Answer: 0 because 2.00E-13 means 2.00×10^{-13} which practically means 0.

4. $i^{192} + i^{193} + i^{194} + i^{195} = ?$

Advanced Calculation and Graphing Techniques with the TI – 83 Plus Graphing Calculator

Unauthorized copying or reuse of any part of this page is illegal.

Solution:

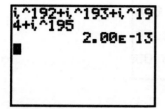

Answer: 0

5. What is the reciprocal of 3+4i

Solution:

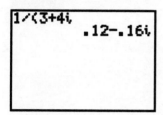

Answer: 0.12 – 0.16i

6. $\dfrac{1+i}{6i+8} = ?$

Solution:

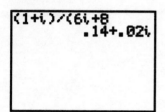

Answer: 0.14 + 0.02i

7. $z = 3\text{cis}\dfrac{\pi}{8}$

$z^3 = ?$

Solution:

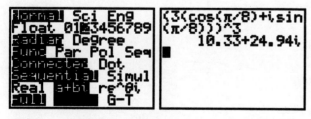

Answer: 10.33+24.94i

8. $z = \dfrac{1+i\sqrt{3}}{-1+i\sqrt{3}}$, what is the value of z in trigonometric form?

Solution:

Advanced Calculation and Graphing Techniques with the TI – 83 Plus Graphing Calculator

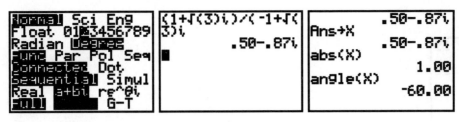

Answer: 1*(cos(-60°)+isin(-60°))

9. A= 3cis40°

   B=4(cos50°+isin50°)

   A.B=?

   Solution:

   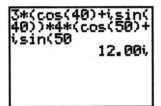

   Answer: 12i

Advanced Calculation and Graphing Techniques with the TI – 83 Plus Graphing Calculator

5.22 Permutations and Combinations

> $n! = n \cdot (n-1) \cdot (n-2) \cdot (n-3) \ldots 3 \cdot 2 \cdot 1$
>
> $P(n,r)$: Number of permutations of r elements chosen from n elements; $P(n,r) = \dfrac{n!}{(n-r)!}$ and
>
> $C(n,r)$: Number of combinations of r elements chosen from n elements; $C(n,r) = \dfrac{n!}{(n-r)! \cdot r!}$
>
> where $r \leq n$. The required functions can be found at the **MATH PRB** menu:

1. $_5P_2 + {}^6P_3 + P(5,3) = ?$

   Solution:

   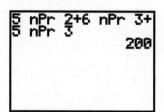

   Answer: 200

2. $\dbinom{5}{3} + C_2^8 + {}_6C_3 = ?$

   Solution:

   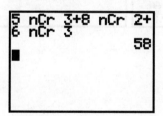

   Answer: 58

3. $\dfrac{(6+3)!}{6! + 3!}$

   Solution:

   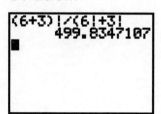

   Answer: 499.83

5.23 Miscellaneous Calculations

Please review Chapter 1 as a whole before starting to do this section. It may also be wise to consult Book 2 necessary.

1. $\cos(2 \sin^{-1}(\frac{-5}{13})) = ?$

    Solution:

    Answer: 0.70

2. $f(x) = \sin(\cos(x))$

    $f(320°) = ?$

    Solution:

     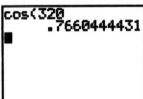  

    Answer: 0.693

3. $f(x) = 14x^2$

    $g(x) = f(\cos x) + f(\tan x)$

    $g(12°) = ?$

    Solution:

    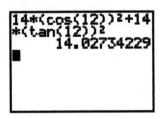

    Mode: Degrees

    Answer: 14.03

Advanced Calculation and Graphing Techniques with the TI – 83 Plus Graphing Calculator

4. cos 30. cos 30° – sin45.sin45°=?

   Solution:

   ```
   cos(30)*cos(π/6)
   -sin(45)*sin(π/4
               -.4680939782
   ```

   Please note that $\pi=180°$

   Answer: - 0.47

5. $f(r,\theta) = r \cos\theta$

   f(12,13)= ?

   Solution:

   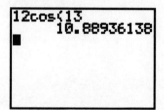

   Mode: Radians

   Answer: 10.89

6. $\csc(\tan^{-1}\frac{1}{\sqrt{2}})= ?$

   Solution:

   ```
   1/sin(tan⁻¹(1/√(2
   )))
               1.732050808
   ```

   Answer: 1.73

7. Arcsin0.6 + $\cos^{-1}$ 0.6=?

   Solution:

   ```
   sin⁻¹(0.6)+cos⁻¹(0
   .6)
               1.570796327
   ```

   Answer: 1.57

Advanced Calculation and Graphing Techniques with the TI – 83 Plus Graphing Calculator

8. $\cos x = \dfrac{1}{3} \Rightarrow \sqrt{\sec x} = ?$

   Solution:

   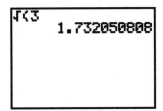

   Answer: 1.73

9. $\left. \begin{array}{l} f(x,y) = \tan(x) + \tan(y) \\ g(x,y) = 1 - \tan(x)\tan(y) \end{array} \right\} \dfrac{f(10°, 20°)}{g(10°, 20°)} = ?$

   Solution:

   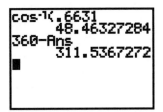

   Answer: 0.577

10. $\cos A = 0.6631$

    $\tan A = -1.12884$

    $\Rightarrow A = ?$ (in degrees)

    Solution:

    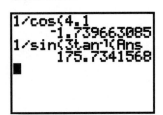

    Answer: 311.54°

11. $\sec 4.1 = x$

    $\csc(3 \operatorname{Arctan} x) = ?$

    Solution:

    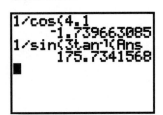

    Mode: Radians

    Answer: 175.73

Advanced Calculation and Graphing Techniques with the TI – 83 Plus Graphing Calculator

12. $\sin A = \dfrac{5}{13}$ $90° \leq A \leq 180°$

    $\cos B = \dfrac{4}{9}$, B is in 4'th quadrant.

    $\sin(A+B) = ?$

    Solution:

    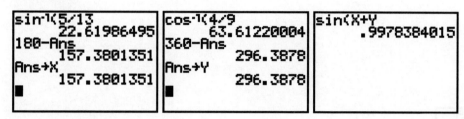

    Mode: Degrees

    Answer: 0.9978

13. In which quadrant is the angle represented by $\text{Arcsin}\left(\dfrac{-3}{5}\right) + \text{Arccos}\left(\dfrac{-12}{13}\right)$?

    Solution:

    Mode: Degrees

    Answer: 2nd quadrant

14. $A \# B = \sqrt{\cos A + \sec B}$

    $4 \# 5 = ?$

    Solution:

    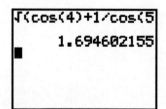

    Mode: Radians

    Answer: 1.69

15. $\dfrac{\cos 15°}{\sin 75°} = ?$

    Solution:

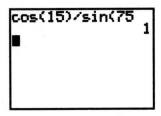

Mode: Degrees

Answer: 1

16. $\dfrac{\tan 25°}{\cos 65°} = ?$

Solution:

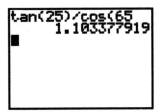

Mode: Degrees

Answer: 1.10

17. $\cos(\tan^{-1}\dfrac{1}{3}) = ?$

Solution:

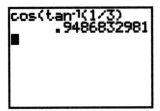

Answer: 0.95

18. $\dfrac{\sin 135° \cdot \cos\dfrac{5\pi}{6}}{\tan 225°} = ?$

Solution:

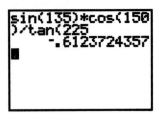

Mode: Degrees

Please note that $\pi = 180°$

Answer: −0.61

19. $\csc\theta = \dfrac{4}{3}$

    $\cos\theta < 0$

    $\tan\theta = ?$

    Solution:

    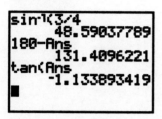

    Mode: Degrees

    Answer: -1.134

20. $\cos 310° + \sin 140° = 2x$

    $\Rightarrow x = ?$

    Solution:

    Mode: Degrees

    Answer: 0.64

21. $\cos\pi - \sin 930° - \csc\left(\dfrac{-5\pi}{2}\right) + \sec(0°) = ?$

    Solution:

    Mode: Degrees

    Please note that $\pi = 180°$

    Answer: 1.5

22. $\tan(-135°) + \cot\left(\dfrac{-7\pi}{8}\right) = ?$

    Solution:

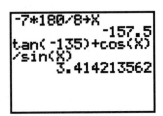

Mode: Degrees

Please note that π = 180°

Answer: 3.41

23. tan (-30°) = - cot x ⇒ x = ? (x is in 3rd quadrant)

   Solution:

   tan (-30°) + cot x = 0

   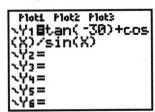 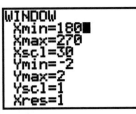 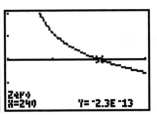

   Mode: Degrees

   Answer: 240

24. cos 210° = ?

   Solution:

   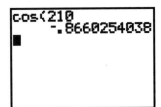

   Mode: Degrees

   Answer: - 0.87

25. sin 330° = ?

   Solution:

   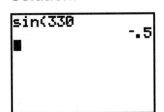

   Mode: Degrees

   Answer: - 0.5

26. $\sec \dfrac{7\pi}{6} \cdot \tan \dfrac{3\pi}{4} \cdot \sin \dfrac{2\pi}{3} = ?$

Solution:

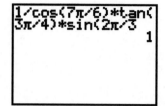

Answer: 1

27. A is in 3rd quadrant and tan A= $\dfrac{8}{15}$

   B is in 2nd quadrant and tan B= $\dfrac{-3}{4}$

   In which quadrant does (A+B) lie?

   Solution:

   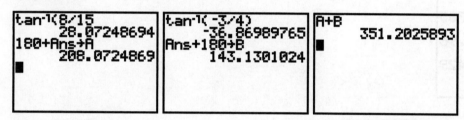

   Mode: Degrees

   Answer: 4th quadrant

28. Given that tanθ= $\dfrac{-5}{12}$ and sinθ is positive.

   cos (2θ) − sin (180°− θ)= ?

   Solution:

   Mode: Degrees

   Answer: 0.320

29. A and B are acute angles and

   tan (A)= $\dfrac{12}{5}$ and sin (B)= $\dfrac{4}{5}$ ⇒ cos (2A + B)= ?

Solution:

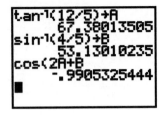

Answer: -0.99

30. $\cos\theta = \dfrac{4}{5}$

    $\sin(2\theta) = ?$

    $\tan(2\theta) = ?$

    Solution:

    Mode: Degrees

    Answer: $\sin(2\theta) = \pm 0.96$ and $\tan(2\theta) = \pm 3.43$

31. $\dfrac{\tan 100° + \tan 35°}{1 - \tan 100° \cdot \tan 35°} = ?$

    Solution:

    Mode: Degrees

    Answer: -1

32. $\tan\theta = \dfrac{3}{4} \Rightarrow \sin\theta = ?$

    Solution:

    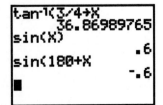

Mode: Degrees

Answer: ±0.6

33. $\cos^2 20° - \sin^2 20° = ?$

    Solution:

    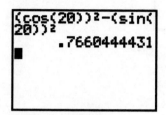

    Mode: Degrees

    Answer: 0.77

34. $\sin A = \dfrac{3}{5}$ and $\cos A < 0$

    $\tan(2A) = ?$

    Solution:

    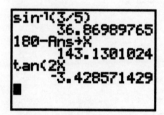

    Mode: Degrees

    Answer: -3.43

35. Important Note: The answers for the following questions are given in both Degree and Radian values as both of them may be asked in the test. It should be noted, however, that these are not two separate answers, but different representations of the same correct answer.

    a. $\sin^{-1} \dfrac{1}{2} = ?$

       Solution:

       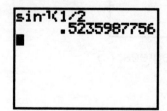   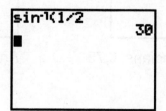

       Answer in Radians: 0.52   Answer in Degrees: 30°

    b. $\sin^{-1} \dfrac{-\sqrt{3}}{2} = ?$

Solution:

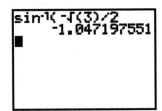
Answer in Radians: -1.05

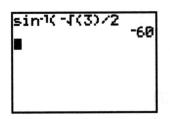
Answer in Degrees: -60°

c. $\cos^{-1}\dfrac{\sqrt{3}}{2}=?$

Solution:

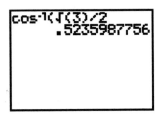
Answer in Radians: 0.52

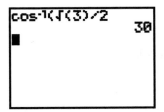
Answer in Degrees: 30°

d. $\text{Arccos}\dfrac{-\sqrt{3}}{2}=?$

Solution:

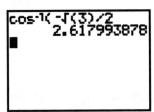
Answer in Radians: 2.62

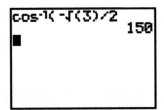
Answer in Degrees: 150°

e. $\tan^{-1}(1)=?$

Solution:

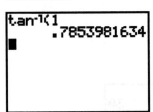
Answer in Radians: 0.79

Answer in Degrees: 45°

f. Arctan(-1)= ?

Solution:

Advanced Calculation and Graphing Techniques with the TI – 83 Plus Graphing Calculator

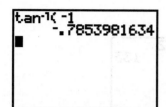 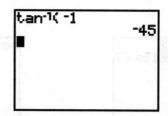

Answer in Radians: - 0.79   Answer in Degrees: -45°

g. $\cot^{-1}(\dfrac{1}{\sqrt{3}})=?$

Solution:

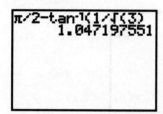 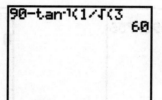

Answer in Radians: 1.05   Answer in Degrees: 60°

h. $\cot^{-1}(\dfrac{-1}{\sqrt{3}})=?$

Solution:

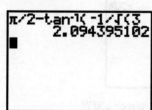 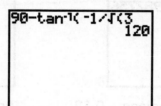

Answer in Radians: 2.09   Answer in Degrees: 120°

i. $\sec^{-1}(\sqrt{2})=?$

Solution:

 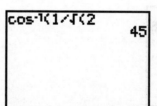

Answer in Radians: 0.79   Answer in Degrees: 45°

j. $\sec^{-1}(-\sqrt{2})=?$

Solution:

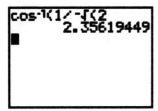

Answer in Radians: 2.36

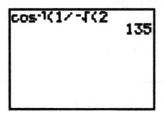

Answer in Degrees: 135°

k. $\csc^{-1}(2)= ?$

Solution:

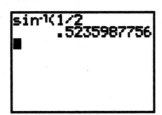

Answer in Radians: 0.52

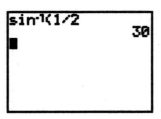

Answer in Degrees: 30°

l. $\csc^{-1}(-2)=?$

Solution:

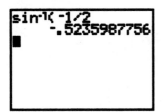

Answer in Radians: -0.52

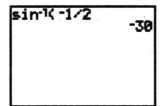

Answer in Degrees: -30°

36. $\sin(\text{Arccos}\frac{4}{5})= ?$

Solution:

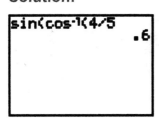

Answer: 0.6

37. $\cos(\text{Arcsin}\frac{-4}{5} + \text{Arccos}\frac{12}{13})= ?$

Solution:

Answer: 0.86

38. $\cos(\sin^{-1}(\frac{-1}{2}))=?$

Solution:

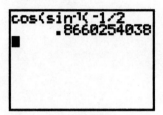

Answer: 0.87

39. $f(x)= \sin x$

   $f^{-1}(\frac{3\pi}{14})= ?$

Solution:

Answer: 0.74

40. $\sin(2\arctan(\frac{-15}{8}))=?$

Solution:

Answer: -0.83

41. $\tan(\text{Arccos } \frac{-3}{5})= ?$

Solution:

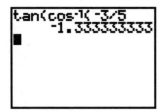

Answer: -1.33

42. a. Arccos(cos $\frac{7\pi}{6}$) = ?

    Solution:

    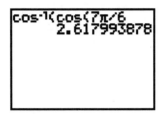

    Mode: Radians

    Answer: 2.62

    b. Arctan(tan $\frac{\pi}{4}$) = ?

    Solution:

    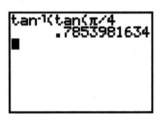

    Mode: Radians

    Answer: 0.79

    c. Arctan(tan $\frac{5\pi}{4}$) = ?

    Solution:

    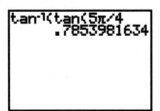

    Mode: Radians

    Answer: 0.79

43. $f(x) = e^x$

    $g(x) = \cos x$

$(f \circ g)(\sqrt{3}) = ?$

Solution:

Mode: Radians

Answer: 0.85

44. $f(x) = \sqrt{2x-2}$

$g(x) = \cos x$

$g^{-1}(f(\sqrt{2})) = ?$

Solution:

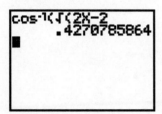

Mode: Radians

Answer: 0.427

45. $\sqrt{2002.2003 - 2001.2002} = ?$

Solution:

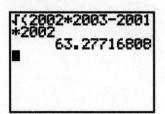

Answer: 63.28

46. $f(x,y) = 2x^2 - y^2$

$g(x) = 5^x$

$g(f(4,3)) = ?$

Solution:

Answer: $1.2 \cdot 10^{16}$

47. $f(x,y) = \sqrt{3x^2 - 4y}$

    $g(x) = 3^x$

    $g(f(2,1)) = ?$

    Solution:

    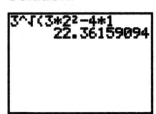

    Answer: 22.36

48. $f(x) = \sqrt{3x - 4}$

    $g(x) = x^3 + x + 1$

    $f(g(2)) = ?$

    Solution:

    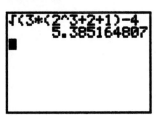

    Answer: 5.39

49. $f(x) = 3x$

    $f(\log_7 4) = ?$

    Solution:

    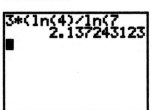

    Answer: 2.14

50. $f(x) = x^2 + 5x - 7$

    $g(x) = x - 5$

    $f(g(\ln 9)) = ?$

Advanced Calculation and Graphing Techniques with the TI – 83 Plus Graphing Calculator

Solution:

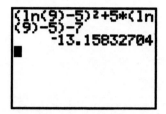

Answer: -13.16

51. $f(x) = \sqrt{x}$

   $g(x) = \sqrt[3]{(x+2)^2}$

   $h(x) = \sqrt[5]{x-4}$

   $h(g(f(2))) = ?$

   Solution:

   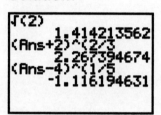

   Answer: -1.12

52. $f(x) = x \ln x$

   $g(x) = 10^{x+1}$

   $g(f(3)) = ?$

   Solution:

   Answer: 19762.27

53. $\sin^{-1}(\cos 200°) = ?$

   Solution:

   ```
   sin⁻¹(cos(200
                -70
   ```

   Mode: Degrees

Answer: -70

54. $\sqrt[3]{y} = 2.6$
$\sqrt[4]{10y} = ?$

Solution:

```
2.6^3
          17.576
Ans→X
          17.576
(10X)^(1/4)
       3.641078238
```

Answer: 3.64

55. $a\Omega b = \dfrac{a}{e + \dfrac{\pi}{b}}$

$(2 \Omega 3) \Omega 4 = ?$

Solution:

```
2/(e+π/3)
       .5311408717
Ans/(e+π/4
       .1515951437
```

Answer: 0.15

56.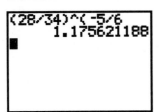

Solution:

```
(28/34)^(-5/6
       1.175621188
```

Answer: 1.18

57. $\sqrt{3} \cdot \sqrt[3]{4} \cdot \sqrt[4]{5} = ?$

Solution:

```
√(3)*4^(1/3)*5^(
1/4)
       4.111400574
```

Advanced Calculation and Graphing Techniques with the TI – 83 Plus Graphing Calculator

Answer: 4.11

58. $a \lozenge b = \dfrac{\sqrt[3]{a} + \sqrt[3]{2b} - 1}{\sqrt{ab+1}}$

    $3 \lozenge \pi = ?$

    Solution:

    Answer: 0.71

59. $\left(-\dfrac{2}{9}\right)^{3/5} = ?$

    Solution:

    Answer: −0.41

60. $f(x) = 3x^2 + 8x - 6$

    Find the negative value of $f^{-1}(0)$

    Solution:

    $f^{-1}(0) = x \Rightarrow f(x) = 0$

    $3x^2 + 8x - 6 = 0$

    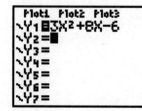 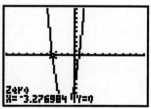

    Answer: −3.28

61. $f(x) = x\sqrt[3]{x}$

    $(f(\sqrt{2}) = ?$

Solution:

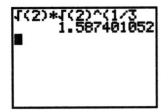

Answer: 1.59

62. $x_0 = \sqrt{2}$

$x_{n+1} = x_n \sqrt[3]{x_n + 2}$

$x_4 = ?$

Solution:

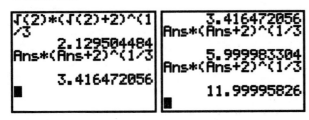

Answer: 12

63. $f(x) = \sqrt[3]{x}$

$g(x) = x^4 + 2$

$(f \circ g)(4) = ?$

Solution:

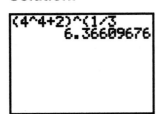

Answer: 6.366

64. $3^{4/3} + 4^{5/4} = ?$

Solution:

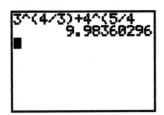

Answer: 9.98

65. $f(x) = |x| + [x]$

$f(1.5) - f(-4.5) = ?$

Advanced Calculation and Graphing Techniques with the TI – 83 Plus Graphing Calculator

Solution:

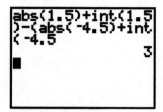

Answer: 3

66. $\log_4(\cos 290°) = ?$

    Solution:

    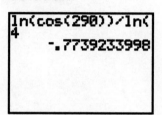

    Mode: Degrees

    Answer: −0.77

67. $\sum_{i=9}^{12} \ln i = ?$

    Solution:

    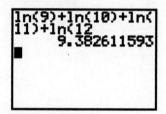

    Answer: 9.38

68. $\log_6 3 = ?$

    Solution:

    Answer: 0.61

69. $\log(\sin 2) + \log(\sin 20) + \log(\sin 20°) = ?$

Solution:

Answer: − 0.55

70. $F(x,y) = \log_y x$

$F(e, \pi^2) = ?$

Solution:

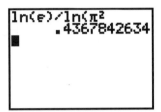

Answer: 0.44

71. $\log_{36} 6 - \log_3 27 + \log_2 (0.25)^{1/3} = ?$

Solution:

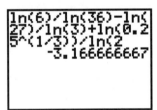

Answer: − 3.17

72. $\log_{\sqrt{5}} 4 - \log_{16} \sqrt{125} = ?$

Solution:

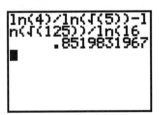

Answer: 0.85

73. $f(x) = x^4 - 88x^3 - 1134x^2 + 3888x + 56135$

$f(99) = ?$

Solution:

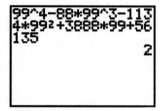

Answer: 2

74. $f(x) = x^3 - 4x^2 + 6x - 4$

$f(3) - f(\sqrt{3}) = ?$

Solution:

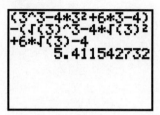

Answer: 5.41

75. What is the remainder when

$3x^4 + 8x^3 + 9x^2 - 3x - 4$ is divided by $x + 1$?

Solution:

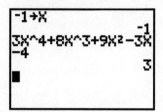

Answer: 3

NOTES

ANSWERS

Advanced Calculation and Graphing Techniques with the TI – 83 Plus Graphing Calculator

5.1 Polynomial Equations

1. Between 0 and 1
2. 2.12
3. NO because P(99) is not zero.
4. Real zeros are -1 and 2 both of which are double.
5. 1
6. -3.226 and 1.189
7. 1 positive and 1 negative real zero.
8. No real zeros, two complex conjugate zeros.
9. 1 positive real zero only.
10. 2.5 or 5/2
11. 1.239-(-2.882)=4.121
12. 2.907+(-0.573)=2.33
13. 1 and -1.333
14. -2.225 or 0.225.
15. -2 + 0 + .667= -1.333
16. -2.637 or 1.137
17. -5.44
18. -1.18+1.18=0
19. Between 1 and 2.
20. -2.449, 2.449, -3.464 or 3.464
21. 1 positive real, 1 negative real and two complex conjugate roots.
22. No
23. 1
24. -7.11 x 10.14= -72.09

5.2 Algebraic Equations

1. -1
2. 0.246
3. 2.294 or 3.228
4. 0.503 or -0.503.
5. -2.158 + 1.158= -1

5.3 Absolute Value Equations

1. 1
2. x= -1.33 or 2

3. x=0.33 or 3
4. No solution
5. 0.6

5.4 Exponential and Logarithmic Equations

1. 6103.5
2. 1.73
3. 7.389+0.693=8.082
4. 4.5
5. 11.66 * 16.24=189.36
6. 0.702
7. 5.129
8. 0.291+3.435=3.726
9. 1.754
10. 1.318

5.5 System of Linear Equations, Matrices and Determinants

1. 1/2
2. $1^2+2^2+3^2=1+4+9=14$
3. (i) $\begin{bmatrix} 2 & 1 \\ -4 & 10 \end{bmatrix}$; (ii) $\begin{bmatrix} 6 & 13 \\ 3 & 5 \end{bmatrix}$; (iii) $\begin{bmatrix} -9 & 11 \\ -15 & 27 \end{bmatrix}$; (iv) $\begin{bmatrix} 2 & -10 \\ -11 & 16 \end{bmatrix}$
4. $A^{-1} = \begin{bmatrix} 0.4 & -0.3 & 0.35 \\ 0.2 & 0.1 & 0.05 \\ 0 & 0 & 0.5 \end{bmatrix}$

5.6 Trigonometric Equations

1. The graph intersects with the x-axis 5 times in the given interval therefore there
2. 39.99°
3. x= 0.312 and tanx= 0.323
4. 70°
5. 0.24 + 0.55 = 0.79
6. 111.47°, 248.53°
7. x has infinitely many values between (0, 2π)
8. 0.41
9. 80.53
10. 234°

11. 1.05
12. 1.57, 3.67, 5.76
13. 22°
14. 240°
15. 15°, 75°, 195°, 255°
16. 15°, 75°, 195°, 255°
17. 11.25° and 56.25°
18. 0.17, 0.70, 1.22
19. 0.52
20. 4.52
21. 135°
22. 1.05, 4.19

5.7 Inverse Trigonometric Equations

1. -0.207 or 1.207
2. 0.92
3. 322.62°

5.8 Polynomial, Algebraic and Absolute Value Inequalities

1. (1,7)
2. (3,4)
3. (0,0.5)
4. [-0.5, 1.84]
5. (-2, 0) or (1,3)
6. (-1, 0) or (2,∞)
7. (-∞, -1] or {0} or [2, ∞)
8. (-∞,0) or (0.67, ∞)
9. (-0.75,1)
10. (0, ∞)
11. (-∞, -4] or [-1, ∞)
12. [1, 3]
13. (3,4)
14. 3rd and 4th quadrants.

5.9 Trigonometric Inequalities

1. (-∞, 0.74)
2. (0, 1.05) or (3.14, 5.24)

3. (0, 0.79) or (3.93, 2π)

4. [120°, 240°] or {0°, 360°}

5.10 Maxima and Minima

1. 1
2. -1
3. (4, -1)
4. 2.08

5.11 Domains and Ranges

1. Because of the even symmetry that the graph has and using the table, we deduce that the domain of the function is as follows: x<-2.73 or x>2.73. This can also be stated as |x|>2.73. Please note that the answer is accurate to the nearest hundredth.

2. Domain: x≠0; Range: y>0.
3. Domain: x≥2; Range: y≤4.
4. The values that must be excluded from the domain are: -1 and 1.
5. y≤9
6. 2kπ<x<(2k+1)π where k is any integer.
7. Domain: x≠-2; Range: y≠3.
8. Domain: x≠1; Range: y≠0.5.
9. Domain: All real numbers except 0 and 2; Range: y≠1.
10. 1≤ y ≤ 10
11. Domain: x ≤ -3 or x ≥ 3; range: y ≥ 0.
12. Domain: -3≤ x ≤ 3; Range: 0≤ y ≤ 3.

5.12 Evenness And Oddness

1. Even (symmetric in the y-axis)
2. Even
3. Odd (symmetric in the origin)
4. Neither odd nor even (no symmetry in the y-axis or the origin)
5. Odd
6. Odd
7. Even
8. Even
9. Neither odd nor even
10. Even

11. Neither odd nor even
12. Even
13. Odd
14. Odd
15. Even
16. Even
17. Odd
18. Neither odd nor even
19. Even
20. Odd
21. Odd
22. Even
23. Even
24. Even
25. Neither odd nor even
26. Neither odd nor even
27. Odd
28. Odd
29. Even
30. Answers are as follows:
    i. Neither odd nor even
    ii. Neither odd nor even
    iii. Even

5.13 Graphs of Trigonometric Functions

1. The function increases throughout in the given interval.
2. Amplitude = (11 + 15) / 2 = 13
   Offset = (11 – 15) / 2 = -2
   Axis of wave: y = -2
3. 2.24
4. 2.45
5. (1.25 – 0.75) / 2 = 0.25
6. 3.93 – 0.79 = 3.14 → π
7. 2.36 – 0.79 = 1.57 → π / 2
8. 1.5

9. (6 − (− 4)) / 2 = 5
10. 0.26
11. 2
12. The function increases between $\pi/4$ and $\pi/2$ and then decreases between $\pi/2$ and $3\pi/4$.
13. 0.79 → $\pi/4$
14. 0.52 → $\pi/6$
15. (2 − 0) / 2 = 1
16. Answers are as follows:
    1. Amplitude = (2 + 2) / 2 = 2; Period = |3.5 − 1.5| = 2; Frequency = 1/2
    2. Amplitude = (0.75 + 0.75) / 2 = 0.75; Period = 4π; Frequency = $1/(4\pi)$
17. 3.14 → π

5.14 Miscellaneous Graphs

1. (3.41, 0.53)
2. 1 point.
3. At x=1 the function is not defined and this is not a vertical asymptote, therefore
4. y-intercept: (0, -8.79), x-intercepts: (-2,0) and (2,0)
5. Answers are as follows:
   a. The inverse is not a function since the graph does not pass the horizontal line test.
   b. The inverse is not a function.
   c. The inverse is not a function.
   d. The inverse is a function.
6. The graph must be shifted 3 units toward right therefore k=3.
7. The graph has the axis of symmetric of x = 0.17.
8. 1.16
9. The vertex is at (1,-5). This is also the maximum point of the graph. The graph doesn't have a minimum point since it is a downward parabola that tends to negative infinity. Equation of the axis of symmetry is x=1 and it is a vertical line.
10. $-6 \leq y \leq 3$

5.15 The Greatest Integer Function

1. 1
2. Min = 0; Max = 1
   Period = | 1 − 0 | = 1 → (The distance between two adjacent maxima or minima)

Advanced Calculation and Graphing Techniques with the TI – 83 Plus Graphing Calculator

      Frequency = 1 → (Frequency = 1 / Period)

      Amplitude = (1 – 0) / 2 = ½ → (Amplitude = (Ymax – Ymin) / 2)

      Offset: (1 + 0) / 2 = ½ → (Offset = (Ymax + Ymin) / 2)

      Axis of wave: y = ½ → (Axis of wave equation is y = Offset)

      Domain: R, Range: $0 \le y \le 1$.

3. The function is not periodic.
4. The range is all integers. The domain is all real numbers.
5. 14

5.16 Parametric Graphs

1. An ellipse
2. A parabola
3. A line
4. Portion of a line that starts at (0,-1)
5. The unit circle
6. A semicircle
7. A quarter circle
8. Portion of a line ranging from (-2,-2) to (2,2)
9. A circle
10. An ellipse (a y-ellipse)
11. A line
12. Portion of a parabola
13. A parabola
14. Portion of a line
15. A parabola
16. Please refer to page 140 solution number 16.

5.17 Polar Graphs

1. 4·3=12
2. 2·4/2=4
3. $\pi \cdot 1.5^2 = 2.25\pi = 7.07$

5.18 Limits

1. -0.75
2. $-\infty$
3. 0

Advanced Calculation and Graphing Techniques with the TI – 83 Plus Graphing Calculator
Unauthorized copying or reuse of any part of this page is illegal.

4. 0
5. 0.25
6. 1.2
7. $+\infty$
8. Limit does not exist because the right hand limit and the left hand limits are not the same.

5.19 Continuity

1. 3
2. 8/3
3. 12

5.20 Horizontal and Vertical Asymptotes

3. Answers are as follows:
   a. Horizontal asymptote: $y=2$
      Vertical asymptotes: $x=-2$ and $x=2$
      Domain: All real numbers except -2 and 2
      Range: $y<2$ and $y \geq 4.5$.
   b. Horizontal asymptote: $y=0$
      Vertical asymptote: $x=2$
      Domain: All real numbers except -2 and 2
      Range: $y \neq 0$ and $y \neq -0.25$.
   c. Horizontal asymptote: $y=1$
      Vertical asymptote: $x=1$
      Domain: All real numbers except -1 and 1
      Range: $y \neq 1$ and $y \neq 3$.
   d. Horizontal asymptote: $y=0$
      Vertical asymptote: $x=3$
      Domain: All real numbers except -3 and 3
      Range: $y>0$ (Although the graph is undefined at $x=-3$, because of its symmetry about the line $x=3$, the y value of 0.0278 exists in the range with the corresponding x value of 9. Therefore the value 0.0278 is not excluded from the range.

2. $x=k/2$ where k is an integer.

5.21 Complex Numbers

1. 0.69 - 3.94i
2. -2+13i

3. 0 because 2.00E-13 means 2.00*10⁻¹³ which practically means 0.
4. 0
5. 0.12 − 0.16i
6. 0.14 + 0.02i
7. 10.33+24.94i
8. 1*(cos(-60°)+isin(-60°))
9. 12i

5.22 Permutations and Combinations

1. 200
2. 58
3. 499.83

5.23 Miscellaneous Calculations

1. 0.70
2. 0.693
3. 14.03
4. − 0.47
5. 10.89
6. 1.73
7. 1.57
8. 1.73
9. 0.577
10. 311.54°
11. 175.73
12. 0.9978
13. 2nd quadrant
14. 1.69
15. 1
16. 1.10
17. 0.95
18. − 0.61
19. -1.134
20. 0.64
21. 1.5
22. 3.41

Advanced Calculation and Graphing Techniques with the TI – 83 Plus Graphing Calculator

Unauthorized copying or reuse of any part of this page is illegal.

23. 240
24. - 0.87
25. - 0.5
26. 1
27. 4th quadrant
28. 0.320
29. -0.99
30. sin (2θ)= ±0.96 and tan (2θ)=±3.43
31. -1
32. ±0.6
33. 0.77
34. -3.43
35. Answers are as follows:
    a. 0.52 or 30°
    b. -1.05 or -60°
    c. 0.52 or 30°
    d. 2.62 or 150°
    e. 0.79 or 45°
    f. - 0.79 or -45°
    g. 1.05 or 60°
    h. 2.09 or 120°
    i. 0.79 or 45°
    j. 2.36 or 135°
    k. 0.52 or 30°
    l. - 0.52 or -30°
36. 0.6
37. 0.86
38. 0.87
39. 0.74
40. - 0.83
41. - 1.33
42. Answers are as follows:
    a. 2.62
    b. 0.79

       c.   0.79
43. 0.85
44. 0.427
45. 63.28
46. $1.2 \cdot 10^{16}$
47. 22.36
48. 5.39
49. 2.14
50. -13.16
51. -1.12
52. 19762.27
53. -70
54. 3.64
55. 0.15
56. 1.18
57. 4.11
58. 0.71
59. - 0.41
60. -3.28
61. 1.59
62. 12
63. 6.366
64. 9.98
65. 3
66. - 0.77
67. 9.38
68. 0.61
69. - 0.55
70. 0.44
71. - 3.17
72. 0.85
73. 2
74. 5.41
75. 3

EMPTY PAGE

Model Test
First Edition

RUSH SAT*II

Mathematics Level IC Subject Test

*SAT is a registered trademark of the College Entrance Examination Board which was not involved in the production of, and does not endorse this product.
Copyright © 2004 by Ruşen MEYLANİ. All Rights Reserved.

RUŞEN MEYLANİ

RUSH
Publications and Educational Consultancy

EMPTY PAGE

Mathematics Level IC Model Test

Test Duration: 60 Minutes

Directions: For each of the following problems, decide which is the BEST of the choices given. If the exact numerical value is not one of the choices, select the choice that best approximates this value. Then fill in the corresponding oval on the answer sheet.

Notes:

- A calculator will be necessary for answering some (but not all) of the questions in this test. For each question you will have to decide whether or not you should use a calculator. The calculator you use must be at least a scientific calculator; programmable calculators and calculators that can display graphs are permitted.

- The only angle measure used on this test is degree measure. Make sure your calculator is in the degree mode.

- Figures that accompany problems in this test are intended to provide information useful in solving the problems. They are drawn as accurately as possible EXCEPT when it is stated in a specific problem that its figure is not drawn to scale. All figures lie in a plane unless otherwise indicated.

- Unless otherwise specified, the domain of any function **f** is assumed to be the set of all real numbers **x** for which **f(x)** is a real number.

Reference Information: The following information is for your reference in answering some of the questions in this test.

- Volume of a right circular cone with radius r and height h: $V = \frac{1}{3}\pi r^2 h$

- Lateral Area of a right circular cone with circumference of the base c and slant height l: $S = \frac{1}{2}cl$

- Volume of a sphere with radius r: $V = \frac{4}{3}\pi r^3$

- Surface Area of sphere with radius r: $S = 4\pi r^2$

- Volume of a pyramid with base area B and height h: $V = \frac{1}{3}Bh$

Unauthorized copying or reuse of any part of this test is illegal.

1C **1C** **1C** **1C** **1C** **1C** **1C** **1C** **1C** **1C** **1C**

Mathematics Level IC Model Test

1. Which of the following illustrates a distributive principle?

A) $(7 + 3) + 5 = 7 + (3 + 5)$

B) $7 + 3 = 3 + 7$

C) $(7 \cdot 3) \cdot 5 = 7 \cdot (3 \cdot 5)$

D) $7 \cdot (3 + 5) = 7 \cdot 3 + 7 \cdot 5$

E) $7 \cdot 3 + 5 = 3 \cdot 7 + 5$

2. In the figure, AB = 10, BC = 2, and D is 4 times as far from A as from C. What is CD?

A) 2

B) 4

C) 6

D) 10

E) 12

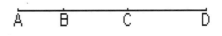

Figure 1

Note: Figure not drawn to scale

3. If n is a positive integer, which of the following is always even?

   I. $n^2 + 3n + 1$

   II. $2n^2 - 4n$

   III. $n^2 - n$

A) I. only

B) II. only

C) III. only

D) II. and III. only

E) I., II. and III.

4. Let A be the set of all numbers a, such that $-3 < a < 7$. Let B be the set of all numbers b such that $5 \leq b < 10$. The intersection, C, of A and B is the set of all numbers c such that

A) $-3 < c \leq 5$

B) $-3 < c < 7$

C) $5 \leq c < 7$

D) $5 \leq c < 10$

E) $-3 < c < 10$

Unauthorized copying or reuse of any part of this test is illegal.

GO ON TO THE NEXT PAGE

Mathematics Level IC Model Test

5. If a > 1 and a^r = 2.4, then a^{-2r}=

A) – 4.8

B) – 5.76

C) $-\dfrac{1}{5.76}$

D) $\dfrac{1}{5.76}$

E) $\dfrac{1}{4.8}$

6. $(x-3)^2 \cdot (x+5) > 0$, if and only if

A) x > – 5 or x ≠ 3

B) x < – 5 and x ≠ 3

C) – 5 < x < 3

D) x > – 5 and x ≠ 3

E) x < – 5

7. A computer is programmed to subtract 5 from M, multiply the result by 4, add 20, and divide the final quantity by 4. The answer given by the computer will be

A) M

B) M – 5

C) 4·M

D) M+20

E) 4·M+20

8. What is the value of $\dfrac{2x-8}{x-8} \cdot \dfrac{x^2-5x-24}{x^2-16}$ where defined?

A) 2 B) x – 4 C) x + 8

D) $\dfrac{2x+6}{x+4}$ E) $\dfrac{2x^2-10x-48}{x^2-16}$

9. If $S+e=\dfrac{S-e}{t}$, $e =$

A) $\dfrac{S\cdot(1+t)}{1-t}$
B) $\dfrac{S\cdot(1-t)}{1+t}$
C) $\dfrac{1-t}{S\cdot(1+t)}$
D) $\dfrac{1+t}{S\cdot(1-t)}$
E) $\dfrac{1+t}{1-t}$

10. What is the value of $\dfrac{1+\frac{4}{x}}{1-\frac{16}{x^2}}\cdot\left(1-\dfrac{4}{x}\right)$ where defined?

A) 0
B) 1
C) x
D) x − 4
E) x + 4

11. If A varies inversely as $\dfrac{1}{B}$ and A = 8 when B = 4; what is A^2 when B = 5?

A) 10
B) 16
C) 25
D) 64
E) 100

12. If x is an acute angle such that $\sin x = \dfrac{4}{y}$ then $\tan x =$

A) $\dfrac{y}{\sqrt{y^2-16}}$
B) $\dfrac{\sqrt{y^2-16}}{y}$
C) $\dfrac{\sqrt{y^2-16}}{4}$
D) $\dfrac{4}{\sqrt{y^2-16}}$
E) $\dfrac{4}{4-y}$

Mathematics Level IC Model Test

13. How many degrees is the angle between the hands of a clock at 7:20?

A) 160°

B) 145°

C) 130°

D) 120°

E) 100°

14. Legs of a right triangle have lengths that are in the ratio of 2:5. If the area of the triangle is 20, what is the length of its hypotenuse?

A) 10

B) $\sqrt{21}$

C) $\sqrt{29}$

D) $2\sqrt{21}$

E) $2\sqrt{29}$

15. If $a = \sqrt{2} - 1$, then $a^2 + 2a + 2$ equals

A) 2

B) 3

C) $\sqrt{2}$

D) $\sqrt{2} - 1$

E) $\sqrt{2} + 1$

16. Point A and circle O are in the same plane. Of all points on circle O, the one that is closest to point A is another point B and length of AB is 8 inches. If the circle has a radius of 8 inches, what is the length, in feet, of the tangent from point A to the circle?

A) 3   B) 12   C) 1

D) 13   E) 5

Mathematics Level IC Model Test

17. In the formula $p = \dfrac{r}{A}$, if $r = 8 \times 10^8$ and $p = 4 \times 10^{-4}$, $A =$

A) 2×10^{-11}

B) 5×10^{11}

C) 2×10^{12}

D) 5×10^{12}

E) 5×10^{12}

18. What is the approximate slope of the line $\sqrt{5}x + 5y - 2\sqrt{5} = 0$?

A) -2.45

B) -2.24

C) -1.24

D) -0.45

E) -0.45

19. How many numbers in the set $\{-5, 0, 4, 8, 15\}$ satisfy the condition $|x - 4| < 8$?

A) none   B) one   C) two   D) three   E) four

20. For what value of x does the function $f(x) = -x^2 - \sqrt{3}x + 4$ become maximum?

A) -3.04

B) -0.87

C) 0

D) 0.87

E) 2.14

21. In $\triangle ABC$, the measure of $\angle B$ is 45° and the measure of $\angle A$ is a. If $|AB|$ is longer than $|AC|$, then

A) 0° < a < 45°   B) 0° < a < 90°   C) 45° < a < 90°

D) 45° < a < 135°   E) 90° < a < 135°

Mathematics Level IC Model Test

22. Four parallel lines are cut by three nonparallel lines. What is the maximum number of points of intersection points that can be obtained?

A) 9

B) 10

C) 11

D) 13

E) 15

23. What is the area in square inches of the largest square that can fit into a rhombus with side of length 20 inches and an interior angle of measure 50°?

A) 141

B) 142

C) 143

D) 285

E) 286

24. If $f(x) = x^2 - 1$ and $g(x) = \sqrt{x} + 4$, what is the approximate value of $g[f(1.5)]$?

A) 5.12

B) 5.50

C) 6.72

D) 18.96

E) 24.72

25. If (x, y) is a point on the function $f(x) = x^2 + x - 3$, for what value(s) of x will y be three times x?

A) no value

B) -1 only

C) 2 only

D) 3 only

E) -1 and 3

Mathematics Level IC Model Test

26. In the semicircle given in figure 2, O is the center of the semicircle, $|AC| = 3$ inches and, $|BC| = 4$ inches. If C is a movable point on arc AB, then the greatest possible area in square inches of $\triangle ACB$ is

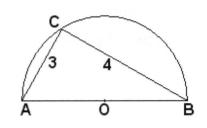

Figure 2
Note: Figure not drawn to scale

A) 50
B) 25
C) 12.5
D) 6.25
E) There is not enough information.

27. A triangle with vertices (-1, -1), (3, -5), and (3, -1) belongs to which of the following classes?

    I.   Right Triangles
    II.   Isosceles Triangles
    III.   Scalene Triangles
    IV.   Equilateral Triangles

A) I. only
B) II. only
C) III. only
D) IV. only
E) I. and II. only

28. The graph given in figure corresponds to which of the following?

A) $y = |x - 2|$
B) $y = |x + 2|$
C) $y = x - 2$
D) $y = x$
E) $y = x + 2$

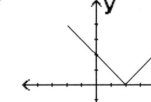

Figure 3

Mathematics Level IC Model Test

29. What is the slope of the perpendicular bisector of segment MT that joins the points M(2,-5) and T(-1,3)?

A) .37
B) .38
C) -3/8
D) 8/3
E) -8/3

30. If the graph of the equation $2x - y + 6 - 2k = 0$ passes through the origin, the value of k is

A) 3
B) 1
C) 0
D) -1
E) -3

31. If $x = -64$, the value of $\frac{1}{4}x^{\frac{2}{3}} + x^{\frac{1}{3}}$ is

A) 0
B) 2
C) 4
D) 8
E) 16

32. If $3^{2x-6} = \left(\frac{1}{9}\right)^x$, x=

A) $\frac{1}{3}$
B) $\frac{1}{9}$
C) $\frac{3}{2}$
D) $\frac{2}{9}$
E) 6

33. $x^2 + y^2 = 25$ represents

A) a circle

B) an ellipse

C) a hyperbola

D) a parabola

E) a straight line

34. The fraction $\dfrac{\sqrt{18} + \sqrt{12}}{\sqrt{6}}$ is equal to

A) $3 + \sqrt{12}$

B) $\sqrt{2} + \sqrt{3}$

C) 5

D) $2\sqrt{3}$

E) $3\sqrt{2}$

35. For what values of A does the equation $x^2 - Ax + A = 0$ have no real roots?

A) $0 \leq A \leq 4$

B) $0 < A < 4$

C) $-4 \leq A \leq 0$

D) $A \leq 0$ or $A \geq 4$

E) $A \leq -4$

36. The equation that expresses the relationship between x and y in the table is

x	1	3	4	8
y	0	4	6	14

A) y = x - 1

B) y = 3x - 3

C) y - 2x + 2 = 0

D) y - 2x - 2 = 0

E) y + 2x + 2 = 0

Mathematics Level IC Model Test

37. The radiator of a car contains 12 quarts of a 25% solution of anti-freeze. If 3 quarts of water are added, what percent of the resulting solution will be anti-freeze?

A) 15%

B) 20%

C) 25%

D) 30%

E) 35%

38. Which of the following expresses the infinite decimal .2161616... as a common fraction?

A) $\dfrac{216}{990}$

B) $\dfrac{214}{999}$

C) $\dfrac{214}{99}$

D) $\dfrac{214}{990}$

E) $\dfrac{216}{1000}$

39. In figure 4, [PA and [PB are tangent to circle O. If angle P measures 80°, how many degrees are in the major arc AB?

A) 60

B) 180

C) 260

D) 100

E) 120

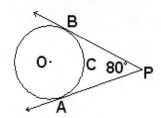

Figure 4

Note: Figure not drawn to scale

Mathematics Level IC Model Test

40. A cubic foot of water is poured into a rectangular tank whose base dimensions are 22 inches by 10 inches. To what height in inches does the water rise?

A) .13
B) .65
C) 1.53
D) 6.11
E) 7.85

41. A bus travels a distance of t miles at 60 mph and returns at 80 mph. What is the average rate of the bus for the round trip?

A) 68.57
B) 70
C) 71.42
D) 72
E) 73

42. If $y = \sqrt{11}x^2 - \sqrt{3}x - \sqrt{2}$, what is the approximate product of the roots?

A) .52
B) -.43
C) -2.35
D) -1.91
E) -.82

43. A circle is inscribed in a triangle with sides 5, 12, and 13. The radius of the circle is

A) 1
B) 1.5
C) 2
D) 2.5
E) 3.5

Mathematics Level IC Model Test

44. Each of the interior angles of a regular polygon measure 160°. How many diagonals does the polygon have?

A) 18

B) 20

C) 40

D) 80

E) 135

45. The solution set of the equation $x - 4\sqrt{x} - 5 = 0$ is

A) {1}

B) {1, 5}

C) {-5}

D) {5}

E) {25}

46. Which of the following is the approximate slope of a line parallel to $\dfrac{x}{.15} - \dfrac{y}{.24} + 3 = 0$

A) -1.60

B) -.63

C) .63

D) 1.60

E) 2.14

47. The graphs of the equations $x^2 + y^2 = 16$ and $y = x^2 - 4$ intersect at how many points?

A) 0

B) 1

C) 2

D) 3

E) 4

48. It is given in figure 5 that in the right triangle ABC, $[DE] \perp [AB]$. If $|BC|=5$, $|AC|=12$, and $|AD|=4$, then $|DE|=?$

A) .67   B) 1   C) 1.67
D) 2   E) 6.5

Figure 5

Note: Figure not drawn to scale

49. Filiz can complete a job in h/3 hours alone and Melissa can complete it in 2h/5 hours alone. If they both work together, how many hours in terms of h will it take them to complete three such jobs?

A) $\dfrac{26h}{15}$ B) $\dfrac{15}{26h}$ C) $\dfrac{6h}{11}$

D) $\dfrac{2h}{11}$ E) $\dfrac{11h}{6}$

50. How many distinct squares are there in figure 6?

A) 25
B) 40
C) 50
D) 55
E) More than 55

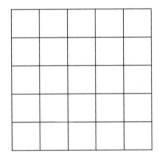

Figure 6

STOP

END OF TEST

Mathematics Level IC Model Test

Answer Sheet

1.	Ⓐ Ⓑ Ⓒ Ⓓ Ⓔ
2.	Ⓐ Ⓑ Ⓒ Ⓓ Ⓔ
3.	Ⓐ Ⓑ Ⓒ Ⓓ Ⓔ
4.	Ⓐ Ⓑ Ⓒ Ⓓ Ⓔ
5.	Ⓐ Ⓑ Ⓒ Ⓓ Ⓔ

6.	Ⓐ Ⓑ Ⓒ Ⓓ Ⓔ
7.	Ⓐ Ⓑ Ⓒ Ⓓ Ⓔ
8.	Ⓐ Ⓑ Ⓒ Ⓓ Ⓔ
9.	Ⓐ Ⓑ Ⓒ Ⓓ Ⓔ
10.	Ⓐ Ⓑ Ⓒ Ⓓ Ⓔ

11.	Ⓐ Ⓑ Ⓒ Ⓓ Ⓔ
12.	Ⓐ Ⓑ Ⓒ Ⓓ Ⓔ
13.	Ⓐ Ⓑ Ⓒ Ⓓ Ⓔ
14.	Ⓐ Ⓑ Ⓒ Ⓓ Ⓔ
15.	Ⓐ Ⓑ Ⓒ Ⓓ Ⓔ

16.	Ⓐ Ⓑ Ⓒ Ⓓ Ⓔ
17.	Ⓐ Ⓑ Ⓒ Ⓓ Ⓔ
18.	Ⓐ Ⓑ Ⓒ Ⓓ Ⓔ
19.	Ⓐ Ⓑ Ⓒ Ⓓ Ⓔ
20.	Ⓐ Ⓑ Ⓒ Ⓓ Ⓔ

21.	Ⓐ Ⓑ Ⓒ Ⓓ Ⓔ
22.	Ⓐ Ⓑ Ⓒ Ⓓ Ⓔ
23.	Ⓐ Ⓑ Ⓒ Ⓓ Ⓔ
24.	Ⓐ Ⓑ Ⓒ Ⓓ Ⓔ
25.	Ⓐ Ⓑ Ⓒ Ⓓ Ⓔ

26.	Ⓐ Ⓑ Ⓒ Ⓓ Ⓔ
27.	Ⓐ Ⓑ Ⓒ Ⓓ Ⓔ
28.	Ⓐ Ⓑ Ⓒ Ⓓ Ⓔ
29.	Ⓐ Ⓑ Ⓒ Ⓓ Ⓔ
30.	Ⓐ Ⓑ Ⓒ Ⓓ Ⓔ

31.	Ⓐ Ⓑ Ⓒ Ⓓ Ⓔ
32.	Ⓐ Ⓑ Ⓒ Ⓓ Ⓔ
33.	Ⓐ Ⓑ Ⓒ Ⓓ Ⓔ
34.	Ⓐ Ⓑ Ⓒ Ⓓ Ⓔ
35.	Ⓐ Ⓑ Ⓒ Ⓓ Ⓔ

36.	Ⓐ Ⓑ Ⓒ Ⓓ Ⓔ
37.	Ⓐ Ⓑ Ⓒ Ⓓ Ⓔ
38.	Ⓐ Ⓑ Ⓒ Ⓓ Ⓔ
39.	Ⓐ Ⓑ Ⓒ Ⓓ Ⓔ
40.	Ⓐ Ⓑ Ⓒ Ⓓ Ⓔ

41.	Ⓐ Ⓑ Ⓒ Ⓓ Ⓔ
42.	Ⓐ Ⓑ Ⓒ Ⓓ Ⓔ
43.	Ⓐ Ⓑ Ⓒ Ⓓ Ⓔ
44.	Ⓐ Ⓑ Ⓒ Ⓓ Ⓔ
45.	Ⓐ Ⓑ Ⓒ Ⓓ Ⓔ

46.	Ⓐ Ⓑ Ⓒ Ⓓ Ⓔ
47.	Ⓐ Ⓑ Ⓒ Ⓓ Ⓔ
48.	Ⓐ Ⓑ Ⓒ Ⓓ Ⓔ
49.	Ⓐ Ⓑ Ⓒ Ⓓ Ⓔ
50.	Ⓐ Ⓑ Ⓒ Ⓓ Ⓔ

Raw Score = Number Correct − Number False / 4

Raw Score = ... − .... / 4 = ...

Unauthorized copying or reuse of any part of this test is illegal.

Mathematics Level IC Model Test

NOTES

Mathematics Level IC Model Test

Answer Key

1.	D
2.	B
3.	D
4.	C
5.	D

6.	D
7.	A
8.	D
9.	B
10.	B

11.	E
12.	D
13.	E
14.	E
15.	B

16.	C
17.	C
18.	E
19.	D
20.	B

21.	B
22.	E
23.	C
24.	A
25.	E

26.	D
27.	E
28.	A
29.	B
30.	A

31.	A
32.	C
33.	A
34.	B
35.	B

36.	C
37.	B
38.	D
39.	C
40.	E

41.	A
42.	B
43.	C
44.	E
45.	E

46.	D
47.	D
48.	C
49.	C
50.	D

Unauthorized copying or reuse of any part of this test is illegal.

Mathematics Level IC Model Test

Scaled Score Conversion Table
Mathematics Level IC Test

Raw Score	Scaled Score	Raw Score	Scaled Score	Raw Score	Scaled Score
50	800	28	590	6	390
49	790	27	580	5	380
48	780	26	570	4	380
47	780	25	560	3	370
46	770	24	550	2	360
45	750	23	540	1	350
44	740	22	530	0	340
43	740	21	520	-1	340
42	730	20	510	-2	330
41	720	19	500	-3	320
40	710	18	490	-4	310
39	710	17	480	-5	300
38	700	16	470	-6	280
37	690	15	460	-7	270
36	680	14	460	-8	260
35	670	13	450	-9	260
34	660	12	440	-10	250
33	650	11	430	-11	240
32	640	10	420		
31	630	9	420		
30	620	8	410		
29	600	7	400		

Model Test
First Edition

RUSH SAT*II

Mathematics Level IIC Subject Test

---

*SAT is a registered trademark of the College Entrance Examination Board which was not involved in the production of, and does not endorse this product.
Copyright © 2004 by Ruşen MEYLANİ. All Rights Reserved.

RUŞEN MEYLANİ

RUSH
Publications and Educational Consultancy

EMPTY PAGE

2C 2C 2C 2C 2C 2C 2C 2C 2C 2C 2C

Mathematics Level IIC Model Test

Test Duration: 60 Minutes

Directions: For each of the following problems, decide which is the BEST of the choices given. If the exact numerical value is not one of the choices, select the choice that best approximates this value. Then fill in the corresponding oval on the answer sheet.

Notes:

- A calculator will be necessary for answering some (but not all) of the questions in this test. For each question you will have to decide whether or not you should use a calculator. The calculator you use must be at least a scientific calculator; programmable calculators and calculators that can display graphs are permitted.

- For some questions in this test you may have to decide whether your calculator should be in the radian mode or the degree mode.

- Figures that accompany problems in this test are intended to provide information useful in solving the problems. They are drawn as accurately as possible EXCEPT when it is stated in a specific problem that its figure is not drawn to scale. All figures lie in a plane unless otherwise indicated.

- Unless otherwise specified, the domain of any function **f** is assumed to be the set of all real numbers **x** for which **f(x)** is a real number.

Reference Information: The following information is for your reference in answering some of the questions in this test.

- Volume of a right circular cone with radius r and height h: $V = \frac{1}{3}\pi r^2 h$

- Lateral Area of a right circular cone with circumference of the base c and slant height l: $S = \frac{1}{2}cl$

- Volume of a sphere with radius r: $V = \frac{4}{3}\pi r^3$

- Surface Area of sphere with radius r: $S = 4\pi r^2$

Volume of a pyramid with base area B and height h: $V = \frac{1}{3}Bh$

Unauthorized copying or reuse of any part of this test is illegal.

GO ON TO THE NEXT PAGE

Mathematics Level IIC Model Test

1. If a and b are distinct prime numbers, which of the following numbers must be odd?

   A) ab
   B) 4a+b
   C) a+b+5
   D) ab-1
   E) 2(a+1)-4b+3

2. If $3x^5=4$, then $5(3x^5)^2=$
   A) 40
   B) 60
   C) 80
   D) 100
   E) None of the above

3. The slopes of two lines that do not intersect can be

   I. equal
   II. reciprocals of each other
   III. negative reciprocals of each other

   A) I only
   B) I and II only
   C) I and III only
   D) II and III only
   E) I, II and III

4. If $x^2 + y^2 = 5$ and $x^2 - y^2 = 3$, then x can be
   A) -1
   B) -2
   C) 3
   D) 4
   E) 5

5. If $\frac{5}{6}x \neq 0$, then $\frac{5}{6}-x$ cannot be

   A) 0
   B) $\frac{5}{6}$
   C) $\frac{6}{5}$
   D) $\frac{25}{36}$
   E) 1

Mathematics Level IIC Model Test

6. $x\left(\dfrac{2}{y} - \dfrac{2}{z}\right) =$

A) $\dfrac{2xz - 2xy}{yz}$ B) $\dfrac{2x}{y-z}$ C) $\dfrac{x}{yz}$

D) $\dfrac{x}{y-z}$ E) $\dfrac{2}{xy - xz}$

7. If $2^{x-1} = A$ and $A = \dfrac{1}{3}$; $x = ?$

A) -1.58   B) -0.58   C) -0.85
D) 0.58   E) 0.85

8. $\log(\sin 20) + \log(\sin 20°) = ?$

A) -0.465   B) -0.466   C) -0.505
D) -0.506   E) -2.681

9. If all variables in the equations xy=35 and yz=45, represent integers less than 1, then what is the maximum value of x+y+z?

A) -9   B) -21   C) -35
D) -45 E) -81

10. If $(-4x)^{2k-1} < 0$, where x is a real number and k is a negative integer, then

A) $x < 0$ B) $x \leq 0$ C) $x > 0$
D) $x \geq 0$ E) x is any real number

Mathematics Level IIC Model Test

11. Given that $A@B = A^B - B^A$ and $2@C = 0$ then C can be

A) -2 B) -1 C) 0
D) 1 E) 2

12. Two quantities A and B are related by the linear regression model given by the following equation: $A = -3.05B + 7.35$. Which of the following can be deduced?

   I. There is a positive correlation between A and B.
   II. When B is less than 20% the predicted minimum integer value of A will be 8.
   III. The slope indicates that as B is increased by 1, A is decreased by 3.05.

(A) II only
(B) I and II only
(C) II and III only
(D) I and III only
(E) I, II, and III

13. Given that $z = t^3 - 1$ and $y = 3t^2 + 4$. If $y - z = 9$ then t can be?

A) -1 B) 0 C) 1 D) 2 E) 3

14. Which of the following relations can represent a function?
   I. $\beta = \{(x,y) | x^2 = y^3; x,y \in R\}$
   II. $\beta = \{(x,y) | x^3 = y^2; x,y \in R\}$
   III. $\beta = \{(x,y) | x^2 + y^2 = 0; x,y \in R\}$

A) I only
B) II only
C) III only
D) I and III only
E) I, II and III

Mathematics Level IIC Model Test

15. If in figure 1 BC is given to be x then which of the following is equal to AC?

A) $x \cdot \sin(\theta)$ B) $x \cdot \sin(\pi - \theta)$

C) $\dfrac{x}{\sin(\theta)}$ D) $\dfrac{x}{\sin(\pi - \theta)}$

E) $\dfrac{x}{\sin(\theta - \pi)}$

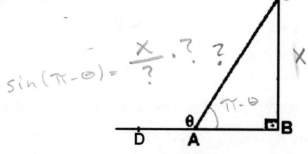

Figure 1

16. $f(x) = \sqrt{3x - 8}$; $g(x) = x^3 + x + 1$; $f(g(2)) = ?$

A) ±5   B) 5   C) 1/5   D) ±1/5

E) none of the above

17. In six years, Murat will be 4/5 of Selin's age. In 15 years, Murat will be 7/8 as old as Selin. If they are both under the age of ten, how old are they now?

A) Murat is 2 and Selin is 4.   B) Murat is 8 and Selin is 6.
C) Murat is 6 and Selin is 9.   D) Murat is 7 and Selin is 9.
E) It cannot be determined from the given information.

18. Statement 1: No Zandies are Mandies;
    Statement 2: All Mandies are Tandies.

Which of the following must be correct?

A) Some Tandies are not Zandies.
B) Some Mandies are Zandies.
C) Some Tandies are Zandies.
D) All Tandies are Mandies.
E) A Tandy that is not a Mandy cannot be a Zandy.

Mathematics Level IIC Model Test

19. sinx = -sinα and cosx = cosα are given. What is x in terms of α?

A) 2π + α   B) 2π - α   C) α - π/2
D) α + π/2   E) π/2 - α

20. Which of the following is the set of all possible x values for the function given by $f(x) = \log_{(5-x)}(x-1)$?

A) x > 1 or x < 5
B) 0 < x < 5 and x ≠ 4
C) 1 < x < 5 or x ≠ 4
D) 1 < x < 5 and x ≠ 4
E) 1 ≤ x ≤ 5 and x ≠ 4

21. Statement: The cube of a number x is less than itself.

The set of all x values that satisfy the above statement are included in the set that contains

A) the real numbers greater than 1.
B) the positive real numbers less than 1.
C) the negative real numbers greater than -1.
D) the real numbers less than -1.
E) the real numbers less than 1.

22. If $f(x) = x^2 - 9$ and f(f(A))=0 then A cannot be

A) -3.5   B) -2.4   C) 0
D) 2.4   E) 3.5

23. If g(x) is given in terms of an arbitrarily selected function f(x) and g(x) = g(-x) then g(x) can be defined as

I. $g(x) = f^2(x)$
II. g(x) = f(x)+f(-x)
III. g(x) = f(-|x|)

A) I only   B) II only   C) I and II only
D) II and III only   E) I, II and III

Mathematics Level IIC Model Test

24. A quadratic equation with integral coefficients has distinct rational roots. Which of the following must be correct?

A) Its discriminant is negative.

B) Its discriminant is nonnegative.

C) Its discriminant is positive definite.

D) Its discriminant is a nonnegative perfect square.

E) Its discriminant is a positive perfect square.

25. $f(x,y) = \sqrt{3x^2 - 4y}$ and $g(x) = 3^x \Rightarrow g(f(2,1)) = ?$

A) 22.3

B) 22.4

C) 12.72

D) 12.73

E) 6561

26. The total cost, in dollars, of a telephone call that is m minutes in length from City R to City M is given by the function $f(m) = 2.34(0.65 \lceil m \rceil + 1)$, where m>0 and $\lceil m \rceil$ is the least integer greater than or equal to m. What is the total cost of a 5.5 minute telephone call from City R to City M?

A) 9 dollars and 94 cents

B) 9 dollars and 95 cents

C) 10 dollars and 71 cents

D) 11 dollars and 46 cents

E) 11 dollars and 47 cents

27. What is the equation of the locus of points at a distance of 10 from the point (3,-1)?

A) $(x-3)^2 + (y+1)^2 = 10$

B) $(x+3)^2 + (y-1)^2 = 10$

C) $(x-3)^2 + (y+1)^2 = 100$

D) $(x+3)^2 + (y-1)^2 = 100$

E) $(x+1)^2 + (y-3)^2 = 100$

Mathematics Level IIC Model Test

28. Base of the right pyramid given in figure 2 is a regular hexagon. If AB=6 inches and BC=4 inches, then what is the volume of the pyramid in cubic inches?

A) $12\sqrt{3}$
B) $24\sqrt{3}$
C) $48\sqrt{3}$
D) $96\sqrt{3}$
E) $192\sqrt{3}$

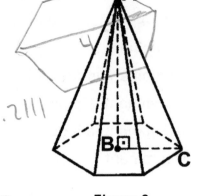

Figure 2
Note: Figure not drawn to scale

29. The first three terms of an arithmetic sequence are 4t, 10t-1, and 12t+2. What is the numerical value of the fiftieth term?

A) 74
B) 79
C) 244
D) 249
E) 254

30. For a polynomial function P(x) what is the remainder when P(x-2) is divided by x+1?

A) P(-3)
B) P(-1)
C) P(3)
D) P(1)
E) None of the above.

31. What is the equation of the line that is equidistant from the points (1,2) and (3,6)?

A) y-4=2(x-2)
B) y-4= -2(x-2)
C) y-4=0.5(x-2)
D) y-4= -0.5(x-2)
E) y-4= -5(x-2)

32. For the trapezoid given in figure 3, AD=4, DC=6 and ∠CBA measures 45°. If the trapezoid is rotated 180° about AB, then what will be the volume of the resulting solid?

A) 368
B) 502
C) 369
D) 503
E) 184

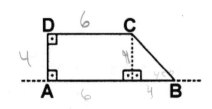

Figure 3
Note: Figure not drawn to scale

Mathematics Level IIC Model Test

33. What is the largest possible domain for the function $f(x) = \sqrt{x+4} + \dfrac{1}{x-3}$?

A) $x > 4$ and $x \neq 3$
B) $x \geq -4$ or $x \neq 3$
C) $x > -4$ and $x \neq 3$
D) $x \geq -4$ or $x \neq 3$
E) $x \geq -4$ and $x \neq 3$

34. The two concentric circles given in figure 4 have radii of 5 and 3 inches and their centre is at O. If measure of angle $\angle AOD$ is $\dfrac{2\pi}{5}$ radians, then what is the area of the shaded region in square inches?

A) $\dfrac{4\pi}{5}$
B) $\dfrac{8\pi}{5}$
C) $\dfrac{12\pi}{5}$
D) $\dfrac{16\pi}{5}$
E) 4π

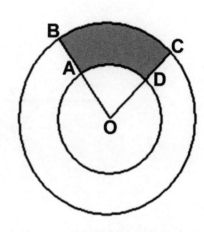

Figure 4
Note: Figure not drawn to scale

35. The point A(3,5) and B(m,n) are symmetric in the line $3x-4y+1=0$. What is the length of segment AB?

A) 2
B) 4
C) 6
D) 8
E) 10

36. What is the obtuse angle between the lines $y = x + 3$ and $y = -\sqrt{3}x + 2$?

A) 45°
B) 75°
C) 105°
D) 120°
E) 135°

37. In a class, there are 23 boys and 19 girls. 10 of the boys and 7 of the girls have participated in the talent show. If a student is selected at random what is the probability that the student has participated in the talent show or he is a boy?

A) 5/7
B) 10/21
C) 10/17
D) 5/21
E) 10/23

Mathematics Level IIC Model Test

For questions 38 and 39, refer to the periodic function given in figure 5 below. Please note that the frequency of this function is $\frac{1}{6}$.

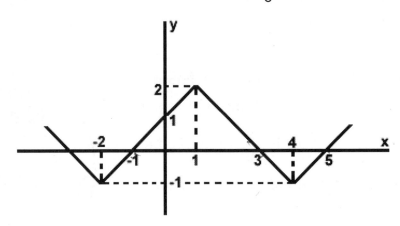

Figure 5

38. What is the amplitude of the function given in figure 5?

A) $\frac{1}{6}$ B) 6.0   C) 3.0

D) 1.5   E) 0.5

39. What is the x-coordinate of the next 10th positive zero of the function given in figure 5?

A) 24   B) 26   C) 27
D) 29   E) 33

40. At a certain harbor, the depth d feet of the water level at time t hours after midnight is modeled by $d(t) = A + B \cdot \sin\left(\frac{\pi}{2} - \frac{\pi t}{6}\right)$, $0 \le t \le 24$, where A, and B are positive constants. P(6, 30) and Q(12, 50) are minimum and maximum points on the graph of d(t) respectively. What is the first time in the 24 hour period when the depth of water is 35 ft?

A) 4 AM   B) 8 AM
C) 4 PM   D) 8 PM
E) None of the above

Mathematics Level IIC Model Test

41. If ln(sin(x))= -1 then what is the least positive value of x?

A) 0.376
B) 2.764
C) 0.377
D) 2.765
E) 21.58

42. In figure 6, f(x) is a continuous function and g(x) has one point of discontinuity. Which of the following is false?

A) (f+g)(x) has at least one zero.
B) (f-g)(x) is always nonnegative.
C) (f·g)(x) has one point of discontinuity.
D) $\left(\dfrac{f}{g}\right)$(x) has one point of discontinuity.
E) The equation f(x)=g(x) has exactly one root.

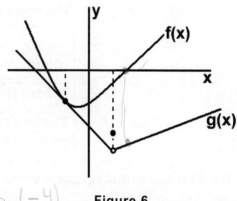

Figure 6

43. What is the real polynomial of the lowest degree whose roots include -2 and 3i where i=$\sqrt{-1}$?

A) $x^3-2x^2-9x-18$
B) $x^3-2x^2+9x+18$
C) $x^3+2x^2-9x+18$
D) $x^3-2x^2+9x-18$
E) $x^3+2x^2+9x+18$

44. If f(x)=x^2-12 then f^{-1}(x)=?

A) x^2-12
B) $\sqrt{x+12}$
C) $-x^2$+12
D) $\pm\sqrt{x+12}$
E) $\dfrac{1}{x^2-12}$

Mathematics Level IIC Model Test

45. $f(x)=2x^2+12x+3$. If the graph of $f(x-k)$ is symmetric about the y axis, what is k?

A) -3
B) 3
C) -15
D) 15
E) None of the above

The sales of cars at the Gofast Car Factory

	Month 1	Month 2	Month 3
Model A	25	32	17
Model B	21	23	19
Model C	19	11	10

46. The table above shows the number of cars that were sold during a three-month sale at the Gofast Car Factory. The prices of models A, B, and C are $59k, $79k, and $99k, respectively, where k denotes thousands of US dollars. Which of the following matrix representations, gives the total income, in k thousands of dollars received from the sale of the cars for each of the three months?

A) $\begin{bmatrix} 25 & 32 & 17 \\ 21 & 23 & 19 \\ 19 & 11 & 10 \end{bmatrix} \begin{bmatrix} 59 & 79 & 99 \end{bmatrix}$

B) $\begin{bmatrix} 59 & 79 & 99 \end{bmatrix} \begin{bmatrix} 25 & 32 & 17 \\ 21 & 23 & 19 \\ 19 & 11 & 10 \end{bmatrix}$

C) $\begin{bmatrix} 25 & 32 & 17 \\ 21 & 23 & 19 \\ 19 & 11 & 10 \end{bmatrix} \begin{bmatrix} 59 \\ 79 \\ 99 \end{bmatrix}$

D) $\begin{bmatrix} 59 \\ 79 \\ 99 \end{bmatrix} \begin{bmatrix} 25 & 32 & 17 \\ 21 & 23 & 19 \\ 19 & 11 & 10 \end{bmatrix}$

E) None of the above

47. Given that f is odd and g is even and that $f(a)=b=g(c)$ then what is the value of $\dfrac{f(-a)}{g(-c)}+f(-a)-g(-c)$?

A) $2b - 1$
B) $-2b + 1$
C) $-2b - 1$
D) -1
E) 1

GO ON TO THE NEXT PAGE

Mathematics Level IIC Model Test

48. If $f(x,y)=x^2-xy+y^2$ for all real numbers x and y, which of the following are correct?

   I. $f(x,y)=f(x,-y)$
   II. $f(x,y)=f(-x,y)$
   III. $f(x,y)=f(-x,-y)$

A) I only B) II only C) III only
D) I and III only E) I, II and III

49. Given two sets of data A={1.0, 2.0, 3.0, 4.0, 5.0} and B={2.0, 2.5, 3.0, 3.5, 4.0}. Which of the following is false?

A) The range of data set A is greater than that of B.
B) Both sets of data have the same mean and median.
C) Each data set contains terms of an arithmetic sequence.
D) Standard deviation of the data set A is less than that of B.
E) A∩B has 8 subsets.

50. The set of points (x,y) given by $x^2-4y^2-4x-24y-32=0$ represents

A) a hyperbola. B) an ellipse. C) a parabola.
D) a circle. E) two lines.

STOP

END OF TEST

Mathematics Level IIC Model Test

NOTES

Mathematics Level IIC Model Test
Answer Sheet

#						#					
1.	Ⓐ	Ⓑ	Ⓒ	Ⓓ	Ⓔ	26.	Ⓐ	Ⓑ	Ⓒ	Ⓓ	Ⓔ
2.	Ⓐ	Ⓑ	Ⓒ	Ⓓ	Ⓔ	27.	Ⓐ	Ⓑ	Ⓒ	Ⓓ	Ⓔ
3.	Ⓐ	Ⓑ	Ⓒ	Ⓓ	Ⓔ	28.	Ⓐ	Ⓑ	Ⓒ	Ⓓ	Ⓔ
4.	Ⓐ	Ⓑ	Ⓒ	Ⓓ	Ⓔ	29.	Ⓐ	Ⓑ	Ⓒ	Ⓓ	Ⓔ
5.	Ⓐ	Ⓑ	Ⓒ	Ⓓ	Ⓔ	30.	Ⓐ	Ⓑ	Ⓒ	Ⓓ	Ⓔ
6.	Ⓐ	Ⓑ	Ⓒ	Ⓓ	Ⓔ	31.	Ⓐ	Ⓑ	Ⓒ	Ⓓ	Ⓔ
7.	Ⓐ	Ⓑ	Ⓒ	Ⓓ	Ⓔ	32.	Ⓐ	Ⓑ	Ⓒ	Ⓓ	Ⓔ
8.	Ⓐ	Ⓑ	Ⓒ	Ⓓ	Ⓔ	33.	Ⓐ	Ⓑ	Ⓒ	Ⓓ	Ⓔ
9.	Ⓐ	Ⓑ	Ⓒ	Ⓓ	Ⓔ	34.	Ⓐ	Ⓑ	Ⓒ	Ⓓ	Ⓔ
10.	Ⓐ	Ⓑ	Ⓒ	Ⓓ	Ⓔ	35.	Ⓐ	Ⓑ	Ⓒ	Ⓓ	Ⓔ
11.	Ⓐ	Ⓑ	Ⓒ	Ⓓ	Ⓔ	36.	Ⓐ	Ⓑ	Ⓒ	Ⓓ	Ⓔ
12.	Ⓐ	Ⓑ	Ⓒ	Ⓓ	Ⓔ	37.	Ⓐ	Ⓑ	Ⓒ	Ⓓ	Ⓔ
13.	Ⓐ	Ⓑ	Ⓒ	Ⓓ	Ⓔ	38.	Ⓐ	Ⓑ	Ⓒ	Ⓓ	Ⓔ
14.	Ⓐ	Ⓑ	Ⓒ	Ⓓ	Ⓔ	39.	Ⓐ	Ⓑ	Ⓒ	Ⓓ	Ⓔ
15.	Ⓐ	Ⓑ	Ⓒ	Ⓓ	Ⓔ	40.	Ⓐ	Ⓑ	Ⓒ	Ⓓ	Ⓔ
16.	Ⓐ	Ⓑ	Ⓒ	Ⓓ	Ⓔ	41.	Ⓐ	Ⓑ	Ⓒ	Ⓓ	Ⓔ
17.	Ⓐ	Ⓑ	Ⓒ	Ⓓ	Ⓔ	42.	Ⓐ	Ⓑ	Ⓒ	Ⓓ	Ⓔ
18.	Ⓐ	Ⓑ	Ⓒ	Ⓓ	Ⓔ	43.	Ⓐ	Ⓑ	Ⓒ	Ⓓ	Ⓔ
19.	Ⓐ	Ⓑ	Ⓒ	Ⓓ	Ⓔ	44.	Ⓐ	Ⓑ	Ⓒ	Ⓓ	Ⓔ
20.	Ⓐ	Ⓑ	Ⓒ	Ⓓ	Ⓔ	45.	Ⓐ	Ⓑ	Ⓒ	Ⓓ	Ⓔ
21.	Ⓐ	Ⓑ	Ⓒ	Ⓓ	Ⓔ	46.	Ⓐ	Ⓑ	Ⓒ	Ⓓ	Ⓔ
22.	Ⓐ	Ⓑ	Ⓒ	Ⓓ	Ⓔ	47.	Ⓐ	Ⓑ	Ⓒ	Ⓓ	Ⓔ
23.	Ⓐ	Ⓑ	Ⓒ	Ⓓ	Ⓔ	48.	Ⓐ	Ⓑ	Ⓒ	Ⓓ	Ⓔ
24.	Ⓐ	Ⓑ	Ⓒ	Ⓓ	Ⓔ	49.	Ⓐ	Ⓑ	Ⓒ	Ⓓ	Ⓔ
25.	Ⓐ	Ⓑ	Ⓒ	Ⓓ	Ⓔ	50.	Ⓐ	Ⓑ	Ⓒ	Ⓓ	Ⓔ

Raw Score = Number Correct − Number False / 4

Raw Score = ……… − ……… / 4 = ………

Unauthorized copying or reuse of any part of this test is illegal.

Mathematics Level IIC Model Test

NOTES

Mathematics Level IIC Model Test

Answer Key

1.	E
2.	C
3.	B
4.	B
5.	B

6.	A
7.	B
8.	D
9.	B
10.	C

11.	E
12.	C
13.	A
14.	D
15.	D

16.	B
17.	C
18.	A
19.	B
20.	D

21.	E
22.	C
23.	D
24.	E
25.	B

26.	E
27.	C
28.	C
29.	D
30.	A

31.	D
32.	E
33.	E
34.	C
35.	B

36.	B
37.	A
38.	D
39.	D
40.	A

41.	C
42.	D
43.	E
44.	D
45.	B

46.	B
47.	C
48.	C
49.	D
50.	E

Unauthorized copying or reuse of any part of this test is illegal.

Mathematics Level IIC Model Test

Scaled Score Conversion Table
Mathematics Level IIC Test

Raw Score	Scaled Score	Raw Score	Scaled Score	Raw Score	Scaled Score
50	800	28	650	6	480
49	800	27	640	5	470
48	800	26	630	4	460
47	800	25	630	3	450
46	800	24	620	2	440
45	800	23	610	1	430
44	800	22	600	0	410
43	800	21	590	-1	390
42	790	20	580	-2	370
41	780	19	570	-3	360
40	770	18	560	-4	340
39	760	17	560	-5	340
38	750	16	550	-6	330
37	740	15	540	-7	320
36	730	14	530	-8	320
35	720	13	530	-9	320
34	710	12	520	-10	320
33	700	11	510	-11	310
32	690	10	500	-12	310
31	680	9	500		
30	670	8	490		
29	660	7	480		

INDEX

π (pi)	18
Absolute Value Equations	57, 70, 100
Accessing a Previous Entry	12
Algebraic Equations	70, 98
Amplitude	63
ANS Variable	12
Asymptotes, Horizontal and Vertical	80
Axis of wave equation	63
CALC intersect	43
CALC maximum	41
CALC Menu	35
CALC minimum	40
CALC value	36
CALC zero	37
Catalog	10
Circle	47
Combinations	18, 81, 177
Complex Numbers	65, 80, 174
Complex Numbers, Operations on	15
Conics, Graphing of	47
Continuity	64, 80, 168
Cube Roots	14
Decimal to Fraction Conversion	14
Default Settings, Restoring	9
Degree	17
Determinants	20, 106
Domains	74, 134
Domains and Ranges, finding	62
e (Euler's constant)	18
Ellipse	49
Evenness	62, 75, 138
Existence of Limit	64
Exponential and Logarithmic Equations	57, 71
Exponential Equations	102
Expressions, Editing	12
Factorial Notation	18
Fractional Powers	14
Frequency	63
Functions, Additional	17
Functions, built in	16
Functions, Composition on, Operations on, Transformations on	32

Advanced Calculation and Graphing Techniques with the TI – 83 Plus Graphing Calculator

Graph Style Icons in the Y= Editor	29
Graphs of Trigonometric Functions	63, 146
Greatest Integer Function	63, 155
Hole	65
Horizontal and Vertical Asymptotes	64, 170
Horizontal asymptote	65
Hyperbola	50
Inequalities, Polynomial, Algebraic and Absolute Value	73, 122
Inequalities, solving	60
Inverse Trigonometric Equations	59, 73, 120
Limits	64, 79, 165
Logarithmic Equations	102
Math Menu	10
Matrices	20, 106
Maxima	74, 132
Maxima and Minima, finding	62
Memory, Resetting	9
Minima	74, 132
Minus Sign, Operational	13
Minus Sign, the Number Minus Sign	13
Miscellaneous Calculations	81, 178
Miscellaneous Graphs	77, 151
n'th Roots	14
Number of Floating (Decimal) Points to be Displayed	13
Oddness	62, 75, 138
Offset	63
Parabola	52
Parametric Graphing	45, 64, 78, 158
PEMDAS is observed with the TI	11
Period	63
Permutations	18, 81, 177
Permutations and Combinations	65
Piecewise Functions, Graphing	31
Polar Graphing	47, 64, 79, 163
Polynomial Equations	69, 91
Polynomial or Algebraic Equations, solving	57
Polynomial Root Finder	25
Radian	17
Ranges	74, 134
Screen Contrast, Adjusting	9
Sequences	19
Series	19
Simple Programming	24

Advanced Calculation and Graphing Techniques with the TI – 83 Plus Graphing Calculator

Simultaneous Equation Solver	**26**
Square Roots	**14**
Statistics	**22**
Storing Values in a Variable	**13**
System of Linear Equations	**106**
System of Linear Equations, Matrices and Determinants	**71**
System of Linear Equations, solving	**58**
Table	**44**
The Greatest Integer Function	**78**
TI BASICS	**7**
TI GRAPHING PRELIMINARIES	**27**
TI Usage	**63**
Trigonometric Equations	**72, 108**
Trigonometric Equations, solving	**58**
Trigonometric Functions, Graphs of	**76**
Trigonometric Inequalities	**74, 128**
Trigonometric Inequalities, solving	**60**
Turning the TI - 83 Plus On and Off	**9**
Vertical asymptote	**65**
Window Settings, Graph Viewing	**30**
Y= Editor	**29**
ZBox	**33**
Zero	**65**
Zoom Cursor	**33**
Zoom In, Zoom Out	**34**
ZOOM Menu	**33**
ZoomFit	**35**
ZSquare	**35**
ZStandard	**35**